W9-BRJ-180

Culture
+ Technology

PETER LANG
New York • Washington, D.C./Baltimore • Bern
Frankfurt am Main • Berlin • Brussels • Vienna • Oxford

Jennifer Daryl Slack & J. Macgregor Wise

Culture
+ Technology

A PRIMER

PETER LANG
New York • Washington, D.C./Baltimore • Bern
Frankfurt am Main • Berlin • Brussels • Vienna • Oxford

Library of Congress Cataloging-in-Publication Data

Slack, Jennifer Daryl.
Culture and technology: a primer / Jennifer Daryl Slack and J. Macgregor Wise.
p. cm.
Includes bibliographical references and index.
1. Technology—Philosophy. 2. Technology—Social aspects.
3. Technology and civilization. I. Wise, J. Macgregor (John Macgregor) II. Title.
T14.S58   306.4'6—dc22   2005015547
ISBN 0-8204-5007-3

Bibliographic information published by **Die Deutsche Bibliothek**.
Die Deutsche Bibliothek lists this publication in the "Deutsche
Nationalbibliografie"; detailed bibliographic data is available
on the Internet at http://dnb.ddb.de/.

Cover art by jd slack

Cover design by Lisa Barfield

Cover photography by M. J. Shupe Photography/
The Studio Gallery Houghton, MI, 906.482.5100

The paper in this book meets the guidelines for permanence and durability
of the Committee on Production Guidelines for Book Longevity
of the Council of Library Resources.

© 2005 Peter Lang Publishing, Inc., New York
275 Seventh Avenue, 28th Floor, New York, NY 10001
www.peterlangusa.com

All rights reserved.
Reprint or reproduction, even partially, in all forms such as microfilm,
xerography, microfiche, microcard, and offset strictly prohibited.

Printed in the United States of America

For Squeaker and the Laddies

Industry
Linocut attributed to James Belvedito, 1933
Library of Congress
Harmon Foundation Inc., Records

Contents

Illustrations

Acknowledgments

THIS BOOK HAS BEEN NURTURED OVER MANY YEARS with the help of numerous people. We have many people to thank jointly and many to thank separately for all manner of assistance and support.

Together we would like to acknowledge the moral and intellectual support of the Conjunctures Group, with whom we have met regularly over several years. You always inspire us. We owe a special debt to Kim Sawchuk for her willingness to read and comment on drafts of this book. Thank you to Leah Lievrouw and Sonia Livingstone for allowing us to test-drive this collaboration as a chapter in *The New Media Handbook*. And though we have only met briefly, Langdon Winner has been a significant inspiration to both of us. We thank Michigan Technological University for the faculty scholarship awarded to Jennifer to facilitate travel from the wilds of Upper Michigan to work with Greg in Phoenix. We are grateful to all those at Peter Lang Publishing who have worked with us on this project: Sophy Craze, Chris Myers, and Damon Zucca. This book has been a wonderful journey, which includes not only the text itself and its ideas but a tagged swan spotted on the shore of Lake Superior, hikes in the Arizona mountain preserves, a search for boots in Phoenix, breakfast in Montreal, and a journey to admire art in Sedona and Jerome.

Jennifer would like to thank the undergraduate students in her classes where these ideas have been worked out over the years: the students in "Technology and Man" at the University of Michigan in the early 1980s; the students in "Communication and Emerging Technologies" at Purdue University in the mid-1980s, and the students in "Philosophy of Technology" at Michigan Technological University, where students have been reading and commenting on duplicated versions of chapters for several years now. A special thanks goes to the graduate students who have learned from and taught me much in "Technology and Discourse" and "Technology, Culture, and Communication." Conversations with Gordon Coonfield were especially pleasurable and inspiring. I deeply appreciate the support of Larry Grossberg and Stuart Hall at crucial moments in my intellectual development. A network of scholars and friends at Michigan Technological University has inspired me to keep working, even when overwhelmed with too many fingers in too many pies. Thanks to Beth Flynn for rock-solid support during those years I now think of as "the worst of times." The late John Flynn for encouraging this book into existence and for continuing to be my technological moral compass. Patty Sotirin for intellectual support and kindness that I treasure beyond words. Daniel Makagon for an optimistic and supportive spirit. Robert Johnson for being the best department chair I ever worked with. Thank you to family and friends for your support and interest. My parents, Pam and Joe Pels. My sisters, Linda Jantzen, Sue Slack, and Laura Pagel. My friends Melvi Grosnick, Phyllis Fredendall, Hannu Leppannen, and Terry Daulton (now you can read it!). Thank you Elise Wise for welcoming me into your home and life when Greg and I were working together in Phoenix. Catherine Wise, sweet little girl that you are, for giving me so much to laugh about when I visited. Finally, I thank Kenny Svenson, husband and partner, for so many things: for a wonderful place to work, for tolerating the inevitable demands that accompany writing a book, and for a life that balances the high-tech hype of most of the rest of the world. And, of course, thank you, Greg. It's been an enjoyable collaboration. I hope it's not our last!

Greg would like to thank the handful of students who signed up for the Special Topics course on "Communication, Culture, and Technology" in spring 1998 at Clemson University, even though the course did not make. It was the process of choosing a text for that class in fall 1997 that led to the extended e-mail conversations with Jennifer, which led to the moment of, "Well, why don't we write one?" Thanks to the students who signed up for it in spring 1999, when it did make. Thanks also to the students in my various "Living Machine on Film" classes at Arizona State University West, who got more of this book in class than they realized. And to my graduate students in spring 2003, who got nearly a draft of the book to work through in class. Thanks to the Humanities Department at Michigan Technological University for inviting me to speak on "Deleuze and Technology" in April 1999. Thanks to my colleagues in the Speech Communication Department at Clemson University (especially Mark Hovind and the late Lynette Eastland) and the Department of Communication Studies at ASU West for giving me such comfortable supportive environments in which to work and learn. I would also like to thank those teachers at the University of Illinois who introduced me to this field of inquiry: James Carey, Cliff Christians, and Andrew Pickering. I especially would like to thank Jennifer for her generosity in letting me write this book with her, for agreeing to visit Phoenix on multiple occasions in the ungodliest heat of summer to work on this book, and for bringing order and coherence to my somewhat chaotic prose. Thanks also to my parents, sister, and in-laws for their support, and for all the extra babysitting. Finally, and most important, my thanks to Elise, for putting up with all this, for letting me buy all those books, and for being my touchstone. Thanks to

Catherine and Brennen, for being too funny and sweet. And thanks to Gabs and Ed for all the loud interruptions.

Jennifer Daryl Slack
J. Macgregor Wise

Potomac Electric Power Company Electric Appliances: Waffle Iron
Photograph by Theodor Horydczak, ca 1920
Library of Congress
Horydczak Collection

Introduction
From Culture and Technology to Technological Culture

W E HAVE WRITTEN THIS BOOK FOR TWO interconnected reasons. First, because we are in the midst of enormous technological and cultural change, technology-related decisions are being made every day with enormous implications for the nature and quality of life on earth. When intelligent people understand the relationship between culture and technology, they can evaluate the options and negotiate better choices. We endeavor to provide ways of comprehending culture and technology that can form the basis for sound understanding and good decision making. Second, we want to keep people from having to reinvent the (theoretical) wheel over and over again. People have been studying the relationship between culture and technology for a very long time. A rich history of debates over technology fills volumes in a variety of fields including: philosophy; history; communication; political science; and science, technology, and society studies. Unfortunately, the excitement generated over the "newness" of some technological developments—for instance, over the role of technology in distance education—seduces us into believing that everything is new and must be studied in a whole new way. We forget that the past has something to teach us.

It is often said that the role of good teaching is simply to save students a lot of time. While it is possible to learn everything anew, it is a lot quicker to gain knowledge of what has already been learned, and to work through what has been tried, rejected, and accepted. It isn't that the learning process holds and reveals the "Truth," but it does provide a strong foundation from which to think through new directions. Frankly, sometimes old ways of understanding the relationship between culture and technology can prove very useful; they can help us avoid a lot of blind alleys, rediscover paths not taken, and thus save a lot of time in an era when time matters so very much. Our second goal, as the title suggests, is to offer a primer—a discussion of the basic concepts, debates, and practices that have been developed—for understanding and acting in what we call "technological culture."

What we present in this book can be considered a series of stories or narratives. We begin with what we call *the received view of culture and technology*. This is the set of commonsense stories our culture tells about technology: stories of progress, convenience, causality, and control. We also tell stories of resistance to the received view: stories of Luddites, Appropriate Technologists, and the Unabomber. Finally, we offer new stories from cultural studies: stories of how the definition of technology shapes technology, how change happens, how agency works, how connections are made and unmade, and stories of space, identity, politics, and globalization.

Stories can be told in many ways, and the stories themselves are influenced by the teller and the circumstances in which the story is told. What we tell here is predominantly a North American story; actually a United States story for the most part, but to some degree a Canadian story as well. These stories may have relevance in other parts of the world, in part because other places have similar stories and because Americans have been quite good at selling their stories elsewhere in the world. Our hope, however, is that this volume will become an impetus for other stories to be told.

Although we draw heavily on a wide variety of fields to make these arguments, our own grounding is cultural studies and communication. Cultural studies is about how inequalities of power are produced, maintained, and transformed through culture. In other words, it argues that culture is a site of struggle and has a role in both reproducing inequality and challenging it. Cultural studies of technology argues that technologies occupy sites of struggle over meanings and power, and that they can both reinforce and undermine structures of inequality.

We come to this project believing that current developments in communication technologies absolutely require all of us to become intelligent participants in technological decision making. But this is not a book about communication technologies alone. Where, after all, in this dawning of the Information Age, does one draw the line regarding types of technologies? If computers are communication technologies, so too are robotics and genetic engineering. It no longer makes sense to think just about radio, television, telephone, and computers as communication technologies, and assembly-line production and heavy industry as industrial technologies. The lines between industrial technology, communication technology, and biology have been blurred beyond meaningful distinction. The concepts, debates, and practices explored in this book can be applied in meaningful ways to any technology you can name, but more importantly, they can be applied to rearticulating the role of technologies in everyday life in what we call "technological culture."

Before we begin, there are four matters that we'd like to consider. The first is a depiction of the central role that technology has played in stories of culture. We counter that centrality by positing a more crucial role for culture. Next we explain what we mean by *culture*. Then, more specifically, we explain what we mean by *technological culture*. Finally, we lay out the structure of our overall argument.

Technology at the Center

Technology has been considered to be a central problem in culture for a very long time. In the familiar stories we tell about what it means to be human, technology more often than not defines the nature of human existence. The evolutionary story of Homo sapiens tells the tale of increasingly sophisticated use of tools. As we've evolved, so have our tools. When we used only crude stone implements, we were one kind of human being living a certain kind of cultural existence. As we developed the wheel, writing technologies, and industrial machines, what it meant to be human and the nature of culture changed. We continue this process of evolution as we develop a human nature and culture based on computers, artificial intelligence, genetic engineering, and nano-technology.

Technology plays three crucial roles in these stories. First, technology is the central defining characteristic of what it means to be human at any particular time. We move through ages: Stone Age, Bronze Age, Iron Age, Industrial Age, Electronic Age, Information/Computer/Digital Age. Each of these "ages" carries with it an image of cultural life that is dramatically different from the others. Second, technology is seen as the causal agent of these ages. The ability to craft stone produces the Stone Age. The development of industrial machines produces the Industrial Age. The computer produces the Information Age. Third, and this point is a bit more subtle, the driving force, or goal, of each of these ages is to perfect these technologies and this stage of development. In this way, technology is the end product, the ultimate effect, and the raison d'être of the age. The raison d'être of the Industrial Age was to take industrial production and culture as far as it could go. The raison d'être of the Information/Computer/Digital Age is to take information/computer technology and culture as far is it can go. These dramatic developments mark out the scope of our evolutionary progress. Cultural theorist James W. Carey sums up all three of these roles of technology in the statement, "Technology is...the central character and actor in our social drama, an end as well as a means." He adds, "at each turn of the historical cycle it appears center stage, in a different guise promising something totally new."[1] Technology plays a central role in defining who we are. Technology shapes our culture; concurrently, culture is organized to give technology its central role. Technology, rather than culture, is at center stage in these stories because in North America technology is like a star-struck actor, stepping on other actors' lines and hogging the limelight. It is not the only actor on the stage, but it is often presented as *a*, if not *the*, central actor.

We would like to flip this formulation around and reposition culture as a more central actor in the technological drama, although perhaps its role is subtler. The technological drama typically unfolds in majestic style with one act following another, featuring waves of innovation, revolution, and change. But in each act, culture is up there on stage: a voice in the discovery of human needs and wants, the methods to meet our needs and wants, the setting and investing in priorities, and in practices of manufacture, distribution, and use.

All these stories are "fictions" in the sense that they are versions of understanding the relationship between culture and technology. It is true that they are powerful stories; they inform our sense of the world and our behavior in it with profound effects. They are versions nonetheless, and thus open to analysis and critique. They are one set of stories among other—and we propose, better—ways of understanding what we call *technological culture*.

Culture as a Whole Way of Life

We take as fundamental Raymond Williams' notions that culture is both "a whole way of life" and "ordinary."[2] For Williams, culture consists of two interrelated processes. First, culture is *tradition*: that is, the meanings, values, and artifacts that are handed down to us, that we learn (and learn about) from families, churches, and schools. These include the works of art and expression that are said to contain the values and worldview of a culture. In addition, culture is *the work of selection*: the selecting, challenging, arranging, and living of these received artifacts and ideas in everyday life. Culture is the process whereby tradition is reconfigured (the term we will introduce later is *rearticulated*) in historical conditions of everyday life and everyday change. The particular shape manifested by the process at a particular point in time is what Williams means by culture as "a whole way of life."

To claim "culture is ordinary" is to acknowledge that these cultural processes occur within the variety of *practices* that constitute everyday life. These include the whole range of activities within which people make meanings in their lives: from everyday expressions and practices such as a conversation over dinner or checking e-mail, to institutional structures and activities such as the structures of education and the practice of designing a technology for public consumption.

Technological Culture

From the perspective of the received view, culture and technology are separate entities. When these are taken to be separable, the task becomes to explain the nature of the relationship between these two. From this perspective, the relevant questions include: Is culture a container from which technology emerges, or into which it is put? How does culture affect technology? How does technology affect culture? This construction and these questions are, for the most part, what North Americans currently understand to be at issue. That is why we call this book *Culture and Technology*.

But if, as we suggest, culture is "a whole way of life"—understood as a process that includes artifacts such as technologies—then technology and culture are not two separate identities whose *relationship* is the central problem of technological culture. From the perspective of culture as a whole way of life, technologies are integral to culture, not separate from it. In that case, it makes less sense to talk about culture *and* technology. Instead, what is needed is a model and a vocabulary that brings technology fully into the concept of culture to begin with. We use the term *technological culture* to mark that difference.

The received view, as you will discover, is a powerful cultural force, and simply replacing the phrase "culture and technology" with "technological culture" is insufficient to overcome the power of its ability to shape thought and practice. One task of this book is to convince you that the concept of technological culture makes better sense as a way to think about issues involving technology. However, once the concept makes sense, you will realize that comfort with the construction "culture and technology" is characteristic of technological culture. So a second task of this book is to encourage you to see, critique, and work to change the dominant role of the received view in our culture.

Consequently, when we write about the received view's take on culture and technology, we tend to use the term "culture and technology." We use the term "technological culture" when we are more specifically encouraging you to take a cultural studies view toward technology and the received view. We trust this will make more sense as you examine the differences between the received view we critique and the cultural studies approach we propose.

The Structure of the Argument

This primer is divided into three sections. In Section I, we map the received view of culture and technology. By "mapping" we mean contextualizing the practices, questions, contradictions, representations, and affects that accompany technological culture. In this section we locate major themes or threads in the way our (primarily US) culture deals with technology. We point to their development in relation to a changing landscape of other cultural forces.

Section II concerns major critical responses to that changing landscape of technological culture. We identify three particularly potent critical responses: Luddism, Appropriate Technology, and the Unabomber. In discussing each of these responses, we consider the ways that they take up the themes discussed in Section I. In doing so, we point to the strengths and weaknesses of each of these responses, and use them to "set the stage" for the intervention we propose in Section III.

Section III explores some of the components of the emerging cultural studies approach to technological culture. It considers the power of language in shaping our assumptions about technology, provides a more nuanced theory of how change happens and how agency works, offers a model of change based on the concepts of articulation and assemblage, and considers the following dimensions that shape technological culture: agency, space, identity, politics, and globalization.

We conclude with some projections for the future of technological culture, and some suggestions for engaging that future with the powerful tools that cultural studies provides.

Chrysler Tank Arsenal Machine Tool
Photograph by Alfred T. Palmer, ca 1940–1946
Library of Congress
Farm Security Administration, Office of War Information
Photography Collection

SECTION I

Culture and Technology
The Received View

CHAPTER ONE

Progress

The Meanings of Progress

IMAGINE IF YOU WILL, standing at the ribbon-cutting ceremony for the launch of a new ocean liner. This is a proud moment: a crowd is cheering, a band is playing. Why is this such an important event? Is it that the ship is the biggest ever, or the most luxurious, or the most sophisticated? Is it that the ship uses the latest instrumentation for guidance, or the most efficient and powerful engines? Perhaps it is all of these things and more. But this ship launching is also an important moment because it is an example of progress. This new machine is evidence that the human race has moved forward, a sign that the race as a whole has improved, that life is now somehow better because this ship is in the world.

Perhaps this example of an ocean liner has reminded you of the story of the Titanic, which in 1912 was the biggest, fastest, most luxurious ship ever launched. It was the unsinkable ship. When the Titanic sank on its maiden voyage, it seemed a slap in the face of progress. Some wondered if we had overreached our place in the world and trusted technology too much.

In US culture, the idea of progress has been closely allied with the idea of technology, and vice versa: technology is progress, just as progress suggests more and new technology. But we have also begun to question this relationship between technology and progress. Is more technology always better? Is the world a better place now than it used to be? The purpose of this chapter is to examine the idea of progress: what progress means, and how technology gets involved. We will look at the story of progress that has been central to the telling of a US story (a story of the Industrial Revolution and the new frontier); and we will also look at how this story is often used as an argument to sell us new technologies, to denigrate other countries and peoples who do not share this story, and to control populations. Finally, we address the fact that it is heresy to question the idea of progress and its relation to technology. Indeed, it is easy to dismiss ideas simply by claiming that they oppose progress, and it is easy to condemn a person simply by saying that he or she is standing in the way of progress.

Defining Progress

The dictionary definition of progress is *to move forward.* If we are walking, we're said to progress down the street. If we're beginning to accomplish a task, we are said to be making progress. When you have read the first chapter of this book, you have made progress. That is, you've done more than before and are on your way to completing the project.

The dictionary meaning is, however, only the beginning of what progress means in an everyday cultural sense. To move forward is to move in one direction: forward as opposed to backward. Consequently, movement forward implies a direction or goal. Similarly, making progress toward the completion of a task implies an endpoint. Progress, then, in its cultural use, is not just movement forward, but *movement toward something*: a goal or endpoint. If a patient is said to be making progress, he or she is moving toward the goal of health. If a disease is said to be progressing, it is advancing, presumably, toward death.

In broad cultural terms, progress is often used to underscore the belief that humankind, as a whole, moves forward. Robert Nisbet, a historian who has written extensively on progress, put it this way:

> Simply stated, the idea of progress holds that mankind has advanced in the past—
> from some aboriginal condition of primitiveness, barbarism, or even nullity—is
> now advancing, and will continue to advance through the foreseeable future.[1]

In addition, as Nisbet sees it, this advancing is not mere movement, but a movement toward something. We aren't marching blindly into a future. Rather, we are advancing toward what we envision as a utopia on Earth. Things will get better and better, and eventually we will achieve what we understand to be "the good life." Progress shows us how far we've come, what we've achieved, and how much better life is now than it used to be. It also reveals to us where we think we are going.

The Goals of Progress

The goals or endpoints of progress are usually unstated, left for the cultural critic to determine by carefully "reading" the culture. However, whether a goal is stated or not, it typically takes the form of what is considered to be "the good life." Most people have a sense of what, for them, the good life entails. It typically involves some of the following: family, community, happiness, leisure, health, wealth, harmony, and so on, though not necessarily in these terms or in this order. Overall, however, two types of goals are associated with progress: material betterment and moral betterment. Material betterment might mean that life is more comfortable, that we are healthier, and that we have more things, more conveniences perhaps, as we discuss in Chapter 2. Moral betterment might mean that spiritually we are more enlightened and that we treat each other better and with more tolerance.

The goals of progress (again, usually assumed as part of unstated cultural knowledge) usually match the fundamental values of a society. Progress at a particular moment in the development of culture could be "a chicken in every pot," indicating a democratic value of universal health and physical well-being; "a car in every garage," indicating the values of widespread personal mobility and private ownership; the absence of war or violence, indicating the value of peace and spiritual enlightenment, or a combination of all three. In any given historical context, understanding the assumed goals of progress is crucial to understanding that culture. Consequently, cultural critics who want to understand technological culture must focus on the everyday practice of culture in order to determine what people think the good life is and the role technologies play in attaining it.

In examining the contemporary relationship between culture and technology, one tendency that has been identified by cultural critics is that people often conflate or collapse the sense of progress as something merely new (a moving forward) with progress as material and moral betterment (a moving toward utopia). For example, using e-mail is said to be progress in relation to the postal service; that is, it is something new, a moving forward. But this newness tends to be equated—without questioning—with the sense of progress as material and moral betterment. Thus, it is assumed, life is better and we are better people with e-mail as part of our cultural experience. When we make this assumption explicit, we can see that the equation is not necessarily true. However, culturally, the tendency to equate the development of new technology with material and moral betterment typically operates without making the assumptions explicit. In part, that is how assumptions gain their power. To interrogate them explicitly is to demystify their power. To facilitate that process and untangle the confusion, we will explore two issues. First, we work to untangle the conflation of newness with material and moral betterment by examining the issue of criteria for measuring progress. Second, we explore the history of the idea of progress as it has come down to us in order to reveal the way this conflation has come about. And, finally, we explore the consequences of this tendency toward conflation.

The Importance of Criteria

How do we know if we are moving forward or not? How do we know if life is better? How do we know if we have progressed? In short, how do we measure progress? Measurement always involves criteria: the standards of reference that allow you to judge. A yardstick is a standard of reference that allows you to determine if one machine is taller than another. But finding an appropriate yardstick to measure progress is especially difficult given the qualitative nature of many of the goals of progress. How does one measure betterment? Happiness? Harmony? Spirituality? Morality?

Because it is much easier to count tangible things, it has become common to use the measure of more *things* as a measure of progress. For example, if we produce more grain than we used to, that's progress. Or, now that more than 50% of US households have a computer, that's progress. Sometimes a measure of more (or fewer) occurrences of something indicates progress. For example, if the mortality rate declines, that is, if fewer people are dying, that must be progress. Or if the Internet has more traffic, that must be progress.

The problem with relying on the numbers of things or occurrences as a measure of progress is that doing so reduces progress to those things that can be counted, losing sight of the qualitative, moral dimensions of progress. Simply put, more is not necessarily better.

For many reasons, more technology and new technology are widely used as the yardsticks for progress. First, because we most often think of technologies as things (as opposed to processes or practices, see Chapter 8), they are easily measured. To the degree that the culture accepts that more things equals progress, more technology is equated unproblematically with progress. Second, technologies in our culture are often linked with key values of the European Enlightenment of the eighteenth century: scientific objectivity, efficiency, and rationality. Because it is a small step from valuing science to valuing its applications, more and new technologies—as applications of science—come to mean progress. Further, technologies allow us to produce more and to undertake new tasks more efficiently, that is, achieving maximum output for minimum work with minimum resources. To the extent that technologies are about achieving efficiency, technology is progress. Finally, technologies are seen as rational. They are about the power of reason and ratiocination to order the world and achieve particular ends. Those ends can be measured; we can chart their progress.

From those criteria considered above, a main criterion for measuring technological progress has been the value of efficiency. A vehicle is more efficient if it goes farther on less fuel. We work more efficiently if it takes us less effort to achieve the same results. The measurement of efficiency often takes the form of a cost/benefit analysis and this is often related to issues of profit. Modern studies of efficiency can be traced back to the work of Frederick Taylor in the early 1900s.[2] Taylor observed, measured, and timed factory workers as they did their tasks, and then worked out ways for them to do their jobs with less effort more quickly and thereby produce more.

The focus on efficiency as the criterion of technological progress has led to complaints of the dehumanization of the workers. Machines are more efficient than humans, so humans are urged to become more machinelike in order to become more efficient. Workers in factories are often taught to perform a task in a particular—efficient—way. They perform a task, and only that task, over and over throughout the day. However, humans are ultimately considered far less efficient than machines: humans require greater and less-predictable energy input in the form of food, rest, entertainment, and so on. Consequently, replacing humans with machines is often seen as embracing efficiency, that is, as progress.

Given the widespread cultural commitments to scientific objectivity, efficiency, rationality, and the ease with which one can see, measure, and count technologies, it is hardly surprising that the mere existence of more and new technology often becomes the only, or primary, yardstick of progress. This leads the culture to focus more on the criteria than the goal. In other words, we assume that the means of achieving progress (technology) is actually the goal itself. We say, "progress equals more technology," not "progress equals the better world created by means of technology."

When technology is seen as the driving force of progress, and this concept is linked to the position that technology shapes culture (a widely held position, as we discuss in Chapter 3), the outcome is a moral imperative on behalf of technology. Technology, and only technology, is what makes the world better. We often hear, "you can't stop progress"; but what is often meant is, "you shouldn't stop progress." To the degree that progress is measured by technology, we are told that we should not stand in the way of technology. We are thus taught to accept things in this culture in the name of progress, even if what we are accepting is harmful to ourselves. David Noble provides the following example:

> A few years ago my mother lost her job to a computer. A legal secretary, she had worked for the same firm for nearly twenty years before being unceremoniously "scrapped" with two days' notice and no pension. The computer created jobs for less-skilled workers and eliminated those of the more-skilled people, like my mother, for whom "retraining" would have meant unlearning. (She was too old to "retool" anyway.) So there she was, home on a Monday morning for the first time in many years, reflecting upon her all-too-familiar plight. She complained about having no job, about the way she was fired after all those years, about the new workers who do not know half of what she knows, about having no pension and the fact that she wasn't getting any younger. But, for all her anger, she was re- signed. Shrugging her shoulders, she repeated to herself as if she had to convince herself, "Well, I guess that's progress."[3]

Progress and technology have become articles of near religious faith held in the heart of North American culture. To question them, to stand in the way of progress and technology, is heresy. We will return to this notion of heresy at the end of this chapter. But first, to fully appreciate how seriously this heresy is taken, we underscore the importance of the story of progress and technology in the development of American culture.

The Story of Progress in American Culture

James Carey and John Quirk wrote, "America was dreamed by Europeans before it was discovered by Columbus."[4] The United States was to be the place where excesses were held in balance: balance between industrial technology and nature, balance between technological betterment and moral betterment, and balance between what has been called "works" (better technology) and "days" (a better life). In this section we chart the development of this story of balance.

By the time Columbus accidentally stumbled upon the Americas, Europe had already had a long and violent history. Civilizations had grown, expanded, and collapsed into darkness again, while new empires had risen. America was seen as a place where civilization could start anew, released from the burden of wars and empires. The Americas were also seen as a new Eden, untouched by the crowding and pollution of European cities. This was a pure place of nature that could redeem Europe. Though the Americas were used primarily as a source of material wealth and resources for war, industry, and empire in Europe, the idea of America as a special place has remained.

When the United States fought for independence, the struggle was seen not only as a political one, as in the creation of a new state, but as a revolution in the conditions of humankind advocating the principles of democracy, freedom, and liberty. Although these principles were echoed in the French Revolution of 1789, the American Revolution was different in that it occurred in the New World. Whereas nature had been exploited and despoiled in Europe, the new country was to embody a balance between nature and the best of what manufacturing technology could offer. Thomas Jefferson wrote, "Let our workshops remain in Europe."[5] Historian Leo Marx famously referred to this balance as the ideal of "the machine in the garden."[6] In the New World, technology and nature would work in harmony.

US leaders such as Jefferson and Benjamin Franklin were not naive, however. They knew that this balance would not happen on its own, and were well aware of the dangers and damages that industrialism could cause.[7] There were those who saw technology and industry as ends in themselves, but Jefferson knew that this viewpoint would upset the balance. He emphasized that technology was a means of achieving progress, not an end in itself. A balance had to be struck between material prosperity as the mark of progress, and moral and spiritual growth as a mark of progress. The nation needed to focus on the goals of progress rather than solely on the means. Franklin, for his part, refused to take out individual patents on his inventions, arguing that the good of society was more important than individual gain.[8]

Unfortunately, the idealism of these founders was diluted. The lure of profit and material wealth became too strong. As the eighteenth century turned into the nineteenth, the Industrial Revolution was heating up. Industry expanded, more goods were produced more cheaply, and soon canals and railways opened the country up to the easy movement of goods and people. Life was prosperous, and the new machines were the most obvious sign of this prosperity. In these times, the view of progress that prevailed was highly technocratic; that is, the adoption of technology was seen as inherently good. Steam engines and railways meant progress in themselves, and the country lost sight of the moral and spiritual dimensions of the term. Ralph Waldo Emerson asked in 1857:

> What have these [mechanic] arts done for the character, for the worth of mankind? Are men better?... 'Tis too plain that with the material power the moral progress has not kept pace. It appears that we have not made a judicious investment. Works and days were offered us, and we took works.[9]

The rapid geographical expansion of the country aided this strong sense of progress, the idea that the United States was constantly moving forward into the future. The frontier experience shaped the character of US culture in crucial ways. As it was seen at the time, civilization strode across the continent, taming nature, the landscape, and the inhabitants with a sense of Manifest Destiny, which is the belief that the continuing expansion of the country across the continent was ordained by God. One of the first, great symbols of this progress was the steam railroad conquering the frontier, a "machine in the garden." Historian Merritt Roe Smith describes a popular allegorical painting of the 1870s titled *American Progress*. The painting depicts a beautiful woman floating across the landscape, a star on her forehead. This figure has been—and still is—used to depict liberty, as she does in the Statue of Liberty, but she was also made to stand for progress. In her right hand is a book; with her left hand she is laying telegraph wire. At her feet are stagecoaches and covered wagons. Behind her follow three railways, and back in the distance, bathed in the rising sun, are an iron bridge and a city. Before her, running away into the darkness, are Native Americans, bear, and buffalo.[10]

Underpinning this vision of Manifest Destiny and progress is evangelical Protestantism. In particular, Calvinism taught the principle of predestination: that there were a chosen few who would inevitably succeed because they had been chosen by God. Applied at a national level, this meant that the United States was God's chosen land, which infused the national character with a fundamental optimism about the future.[11]

The promise of the machine in the garden was tarnished as the nineteenth century progressed, and the pollution, environmental destruction, and slums of Europe were recreated in the New World. In addition, the bloodiness and destruction of the Civil War shook the faith in the country as a place of peace and prosperity. With its brutal war and industrial machines, how could the United States be regarded as the land of progress? In spite of setbacks, however, the notion of technological progress remained strong, largely due to the excitement over yet another new technology as a symbol of progress: electricity.

Unlike the menace of large machines, electricity appeared clean, mysterious, even supernatural. When applied to communication, first in the form of the telegraph, electricity was seen as revolutionizing the country. Prior to the telegraph, communication had been synonymous with transportation. Messages traveled at the speed of horses, carts, ships, or trains. But with the telegraph, one could communicate instantaneously with people hundreds of miles away. One became aware of a sense of simultaneity, the knowledge that others were living their lives at that moment across the nation. The telegraph also made a profound impact on the economy in that it helped to create a national market for goods. Before the telegraph, it was difficult to find out how a crop was doing in Ohio or how production was at a factory in Pennsylvania. The telegraph provided the commodity market with more accurate and immediate information.[12]

Electricity transformed street lights, shop-window displays, department stores, drawing rooms, and thereby the nature of city life, both public and private.[13] Light sources changed from hazardous torches, open flames, or gas lights to relatively safe light bulbs and filaments. As we moved from the "primitiveness" of the open flame to the science of electric light, we experienced progress for which technology was deemed responsible.

Electricity continued to be the symbol of progress through the middle of the twentieth century. The electrification of more and new technologies, in particular household appliances, and the growing availability of electricity to many parts of the nation were taken as evidence of progress. Large-scale projects like the Hoover Dam, completed in 1936, provided electric power to the Southwest. In 1933 President Roosevelt created the Tennessee Valley Authority, a Federal corporation charged to develop the Tennessee River system to promote navigation, flood control, and the production and distribution of electricity to wide regions of the Southeast. Projects such as these can be seen as a continuation of electricity as the primary symbol for progress. Eventually, in the wake of concern over atmospheric pollution caused by coal-burning power plants and the environmental destruction caused by electricity-producing dams, electricity began to wane as a symbol of progress.

Nuclear power, awesome in its own right, replaced electricity as the dominant symbol of progress and continued the tradition. Eventually, nuclear power too revealed a darker side to technological progress in the form of the nuclear bomb and the threat of radioactive contamination.

The most recent symbol of progress is the digital computer, which has dominated the American imagination since the 1950s. The computer differentiates itself from other electrical technologies in that it, unlike technologies that mimic the physical work of humans, mimics the work of the of the mind.[14] The progress implicitly embodied in the digital computer is the ability to process more data and an expansion of the concept of thinking. There are people who argue that computers may someday progress so far beyond human capabilities that we could create— some say have already created—artificial intelligences. Though this possibility is the stuff of fantasies and nightmares, as articulated in any number of science-fiction novels and films, there are those who see the surpassing of the human as a positive development. Humans are hindered from evolving or progressing further, these people argue, by the limitations of the human body. The true goal of human progress is the expansion of the mind according to some; and if we could somehow abandon the body, we would truly evolve.[15] We would become "post-human."

Beyond the more fantastic images of the post-human, digital technologies allow us to progress because, as MIT professor Nicholas Negroponte has put it, what we used to accomplish by physically moving atoms around—for example, shipping books or delivering newspapers—we can now accomplish by sending bits of electronic information instantaneously and cheaply from place to place.[16] Instead of having to work with physical models of cars, buildings, or even bodies, we can create virtual representations of them in computers and submit them to any number of virtual tests and stresses.

Two Concepts That Underpin and Help to Sustain This Story: Evolution and the Sublime

In the stories of these American revolutions, from 1776 to the information revolution of today, technology has played a principle (determining) role in our conceptions of progress, to the extent that we have confused the profusion of technologies with progress. The machines themselves, not the goals of progress, have come to play center stage. This story of progress has been given additional heft because it draws on two other powerful concepts: *evolution* and the *sublime*.

Progress and evolution are often conflated, in part due to a pervasive misunderstanding of the idea of evolution. The misunderstanding asserts that as we evolve we are likewise progressing; that is, we are becoming better, more perfect human beings. In other words, as we evolve toward something, we are progressing into something better. Evolution is thus given a "progressivist" twist in popular accounts. But this is not the intent of the theory of evolution.

Evolution is the slow adaptation of living creatures to environmental conditions over the course of generations. The creatures that do not survive do not pass on their particular genetic attributes to future generations. Groups are selected to survive on the basis of randomly occurring genetic mutations. The idea of natural selection is often oversimplified to the idea of "the survival of the fittest," which purports that surviving generations are stronger, faster, smarter, and more complex. But this is not necessarily the case. In general, the direction of evolution has been from the simple to the complex, from single-celled organisms to multi-celled ones. However, this in no way guarantees the survival of the most complex organisms in the face of changing environmental conditions. Cockroaches are, after all, more fit to survive a nuclear war than humans. So even though less complex organisms might be better adapted to changing environmental conditions, we are unlikely to evolve back into single-celled creatures anytime in the near future. Consequently, evolutionary theory resists the notion that humans are necessarily better or more advanced than other species. We have merely evolved *differently*.

Evolution, in its misunderstood version, underpinned the idea of progress in the nineteenth century and beyond by providing a scientific version of the principle of Manifest Destiny and evangelical Protestantism. According to this version of evolution, it was "natural" that the nation would achieve greatness since it was, as was widely believed, at the forefront of technological development. Further, technological and national "might," linked to this misunderstood idea of evolution, promoted the belief that "might makes right," for only the fittest survive.

The second concept that undergirds progress is the notion of the sublime. The idea of the sublime involves a glimpse of perfection, the sense that one is viewing God or God's work. The sublime is awe inspiring, an overpowering combination of dread and reverence.[17] The United States possesses its share of sublime wonders such as Niagara Falls, the Grand Canyon, and the Rocky Mountains. Leo Marx saw behind his idea of the machine in the garden another type of sublime: the technological sublime. The advance of technology at the time seemed divinely inspired, and people stood in awe of the large steam engines or of electricity itself. The technological sublime refers to the almost religious-like reverence paid to machines. These machines were much more powerful than individual humans and held out the promise of being able to achieve perfection. Whereas hand-made goods have irregularities and imperfections, those made by machines, potentially, do not. The technological sublime, then, carries with it a fear of being overwhelmed, an attraction to the beauty of the perfection of the machine and its products, and, most of all, a reverence for the awesome power of the machine.

The technological sublime that Marx described was what we would call the "mechanical sublime," the divine nature of large, industrial machines. But when the machine began to fail as an untarnished symbol of progress after the Civil War, electricity took on the mantle of the sublime, what James Carey and John Quirk call "the electronic sublime."[18] In contrast to the smoke, soot, and grease of mechanical engines, electricity seemed pure and clean. Electricity is intangible; its nature is almost mystical. People even feared the new electrical telegraph lines that sprung up in the mid-nineteenth century. It was said that when the wind blew over the electric lines they produced an eerie moaning sound, and people went out of their way to avoid them.

More recently, as the symbols of progress have shifted yet again and electricity has become commonplace, our feelings about electric technologies have shifted. For the most part, electricity is now seen as polluting, and nuclear generators as dangerous. Where once we waxed poetic about turbines and railroads, electrical dams, dynamos, and nuclear reactors, our imaginations now soar with effusive paeans to digital technologies, especially as they relate to cyberspace. We are faced with what we may call the "digital sublime."

We discuss these notions of evolution and the sublime here to better understand the power of the story of progress. Why would so many people accept technology as progress without question, even if it hurt them, as it did David Noble's mother in the earlier example? We are persuaded by progress because we are persuaded by the logic (*logos*) of the argument that it is better to be efficient, rational, and scientific. We are also persuaded by the logic and ethic (*ethos*) of the argument of evolution (we trust science and scientists) that progress is inevitable. And finally, we are persuaded by the deeply emotional argument (*pathos*) of the sublime; persuaded by our own feelings of fear, awe, and expectation.

The Uses of the Progress Story

Stories are not neutral. We tell stories to make a point, to educate, to persuade, to entertain. Stories have their uses. It is important to emphasize that what we have covered above is a story, though it might seem like history. Culturally, we are all acclimated to accepting history as the "Truth" about the way things actually happened. But in telling history, one is telling a story. History, like any story, is always told by someone to someone else for a particular purpose. Told by someone else, the story might be different. For example, a Native American version or a Canadian version of the events since the settling of the New World would be different from what we have described. Stories are told for different reasons: to persuade us to go to war, to persuade us to buy a product, to convince us that what our ancestors did was correct and justified, or to make us feel comfortable (or uncomfortable) with our place in the world. We have told the story above as much as possible in the terms in which it is usually related; this does not mean that we agree with this story or the justifications that it provides.

The story of progress as told above has been put to four major uses in the United States: to promote a version of "a better life," to sell us things, to judge others, and to control populations.

Promoting a Better Life

Historian Robert Nisbet argues that the progress story emphasizes that change is good and that change promotes a better life. He believes that as long as we continue to tell, believe in, and live the progress story, our culture will not stagnate. The progress story is essentially a revolutionary story; it promoted, and continues to promote, both political and technological change. Many positive outcomes, services, and products can be attributed to telling, believing, and living the progress story, including democracy, sanitation, education, cyberspace, and life-saving medical advances.[19] The progress story thus promotes a particular version of a better life.

Selling Us Something

Historian Merritt Roe Smith relates that in the mid-nineteenth century the working classes were speaking out against progress because it was being used as an excuse to install new machines in the factories, thus putting them out of work. They understood that it is crucial to ask the questions, Progress for whom? Who benefits? And the answer was that it wasn't them. But at the same time that they were contesting progress in the workplace, they were eagerly buying the new products that were being produced by these new machines. By purchasing these products, not only were these people supporting the country's economy but also actively participating in what they saw as the future. In other words, they could put aside their own individual issues and participate in the broader sense of Manifest Destiny and progress.[20]

Technological progress is often a theme of advertising. We purchase things because they are New! Advanced! Improved! We purchase new computers because they are faster than old computers. We may also purchase a technological object because of its beauty or power. Look at advertisements for cars, stereo equipment, and computers; they are replete with claims of new improvements and awesome appearances.

Interestingly, explicit appeals to progress as a justification for buying seem to be diminishing. In fact, it would seem quaint, or old-fashioned, to defend one's purchase of a new car as "progress." Instead, appeals to progress and the sublime have taken a new form. We are now inclined to purchase technologies, not for a sense of the progress of civilization or for the appreciation of grandeur, but for their contemporary manifestation. The "cool" and the "neat" are what we think of as the new *mini-sublime*. Think about the stores that cater to tantalizing buyers with "cool" or "neat" gadgets: from the high-tech of The Sharper Image to the low-tech pleasures of office-supply stores, cooking-supply stores, and hardware stores. Just think about how often, when presented with a new device, the response is simply this: "Cool!" or "Neat!"

In addition to the appeal of the cool and the neat, a more considered justification for buying is frequently convenience. One might easily defend the purchase of a new car for its conveniences: air conditioning, remote starting, GPS locating, and so on. The ascendance of convenience, the topic of the next chapter, does not mean that progress is becoming passé; rather, it suggests that what constitutes progress has become closely allied with the value of convenience. While progress is still what is more, new, advanced, better, cool, or neat, it is also more convenient.

Judging and Controlling Others

CIVILIZED AND PRIMITIVE: When Western European explorers first encountered the cultures of the Americas, Africa, and the South Seas, they were perplexed. These cultures were so very different from their own. The people had much less technology than the Europeans. Rather than concluding that these others were simply different and leaving it at that, Western Europeans drew on the story of progress to explain the situation.

In the received view, the story of progress presents a linear view of cultural development: It moves from simple to complex, and from less technologically advanced to more technologically advanced. It also concludes that every culture must progress in this way: first because progress is universal, and second because it is divinely inspired. The assumption is that these other cultures must be at an earlier stage on the same line as Western progress. Furthermore, these cultures could be expected to progress in the same way that Europe did until they eventually reached the level of European culture. Therefore, the story goes, they were deemed primitives who would one day be civilized; the criterion by which the progress of their civilization could be measured was, predictably, the technologies they embraced.

The progress narrative, then, was used to label cultures as either civilized or primitive. Those labeled primitive were considered less intelligent, less cultured, and beneath European culture. Colonization of primitives was not merely justified, it was considered a moral responsibility; for with assistance, primitives might be brought into the fold of a better, civilized life. Hand in hand with colonization, labeling cultures primitive or civilized fed into the rise of nationalism on the one hand and on the forced technological development of cultures on the other.

NATIONALISM: A nation is a group of people recognized as having shared characteristics that unify them as a single entity. The group as a whole seems to have a unified identity. Beyond the less formal categories of membership or citizenship, being a member of a nation involves a shared identity and an emotional bond. For example, we may be citizens of the United States, but we might think of our nationality as American. Nationalism is devotion to one's nation, a pride in one's national accomplishments. Two fundamental aspects of the nation are, on the one hand, the recognition that there are thousands if not millions of others with whom you share this identity and, on the other hand, the recognition that there are millions of others who do not. Nation is not only a label indicating membership; it is a means of differentiating *us* from *them*. The progress narrative is easily used as a means to differentiate nations, particularly to denigrate some and elevate others, and levels of technology have become part of the yardstick by which to measure and compare.

The practice of measuring the progress of nations with technology is dramatically illustrated by the great industrial expositions and world's fairs of the nineteenth and early-twentieth centuries. Much like the modern Olympic Games, these events were opportunities for all nations to gather peacefully with an attitude of good will to share in the best of what each nation had to offer. But also like the Olympic Games, there was competition behind the exhibition.

The first major industrial exposition, and the model for those that followed, was the Great Exhibition of the Works of Industry of All Nations, which opened in London in 1851. The Great Exhibition, as it was called, was held in a newly constructed building made entirely of glass and iron, which was referred to as the Crystal Palace. The Crystal Palace was an accomplishment in itself, the first building made almost entirely of prefabricated parts. Each nation was allotted a space to display inventions, innovations, machines, and the products of machines such as textiles and artwork. Because the Great Exhibition was held in London, the British claimed a good portion of the floor space for their products and those of their colonies, such as India. In a didactic move, India was placed at the center of the hall, but the selection and arrangement of the display emphasized the humble nature of the inventions, innovations, machines, and products of India. The Great Exhibition was thus an opportunity for Britain to show off its technological superiority. In addition it was an opportunity to show off its superior cultural character. While Europe before 1851 was characterized by violent revolution, Britain alone was at peace with others and with its own working classes. The Crystal Palace, dubbed in the press as "The Palace of Peace," was meant to fuse the ideas of British national character, moral and cultural superiority, industrial superiority, and progress.[21]

Other international expositions followed: 1853 in New York; 1867 in Paris; the Centennial Exposition in 1876 in Philadelphia; and 1889 in Paris, for which the Eiffel Tower was built. However, the strongest assertion of the technological-progress story occurred at the Chicago World's Fair in 1933. The fair's guidebook stated: "Science discovers, genius invents, industry applies, and man adapts himself to, or is moulded by, new things." It summarized: "Science finds—Industry applies—Man conforms."[22] Across the varied exhibits of the fair were similar statements, reinforcing not only a belief in technological determinism, that is, a belief that technology drives culture, but a belief in technological progress, that is, a belief that technology drives civilization. The superiority of nations had become a matter of fusing technology, progress, national character, and moral character.

DEVELOPMENT: The story of technological progress was not just used for national self-aggrandizement. According to the linear view of progress, these other, primitive cultures would eventually progress or develop to the levels of the industrialized countries. So why not help them along? Working under the assumption that all nations inevitably will become technologized (and want to become technologized), Westerners advanced the idea that these countries could be helped by being given or loaned advanced technologies. More technology would help these nations "leap-frog" over the intervening stages of technological development, contribute to cultural progress, and render them civilized sooner. "Development" is the term that was widely used to describe this process.

The term development has much in common with the term progress. Like progress, development assumes a constant move forward toward some goal. For example, one develops into something: a boy into a man, a kitten into a cat, a pupa into a butterfly. However, the meaning of development carries with it a stronger sense of inevitability than progress. We can label the stages of development—infant, child, adolescent, adult—and be pretty certain that each person will move though these stages toward the inevitable conclusion. When this idea was applied to nations, each was depicted as located at a particular stage of development: some were developed, some were less developed, and some were undeveloped—the so-called Third World. European and North American programs designed to help nations develop were based on the assumption that all less- and undeveloped countries would eventually look like the countries of Europe or North America, and that they would want to. This is an egocentric assumption, at best.

Large development programs were put into place worldwide in the 1950s and 1960s. Industrial technologies, communication technologies, agricultural technologies, as well as electrification projects, dam building, and so on, were put in place with little regard for local cultures or social norms. Some were successful, but many were not. Across the board, however, the most prevalent result of these programs was the plunging of the Third World into incredible debt. In addition, when traditional farming practices were replaced with industrial farming practices that focused on cash crops for export—such as cotton—many countries found it difficult to feed their own people. As a result, these countries became dependent on the West for food, resources, and technical know-how. The progress story thus discriminates among different cultures, promotes a particular version of technological development for those "less civilized," and generates problematic dependencies among nations.

Because of these problems and resulting dependencies, the term development has acquired strong negative connotations. For many, the failings of development result from its top-down approach, where decisions are made by an elite at the top of a nation's social hierarchy, or by a few technical experts from an industrialized country, and then imposed on the rest of the population without their input or consent. More recently, there has been a move to rehabilitate the term development by presenting a grassroots model of development, in which technological and cultural change are instigated at a local level with local input and consent. The grassroots model of development seeks to distance itself from the progress model in that the final shape and character of a nation would be determined internally and not by the external imperative of technology. The grassroots model seeks to do away with the predetermined outcome of development, and substitutes moral or cultural criteria, in addition to technological criteria such as efficiency, to point in a direction of desired development.[23]

POLITICS: We have seen in the discussion above how the technological progress narrative is used in international politics, but it is also used to influence politics within a nation. When people are willing to believe that technology drives progress and that technological change is inevitable and good, people are more willing to accept the advice of the experts, that is, the technologists who claim to know how technological change is accomplished.[24] People become geared to expect and accept technological change. When, in addition, technological progress is seen as inevitable, there is no need to shape or guide science and technology. Major technological decisions become mere technical matters that do not demand or justify the consultation of nonexperts. Consequently, the technological progress story has been used to promote more authoritarian and technocratic decision making and to suppress democratic decision making. We will return to the ideas of technological politics later on in this book, but it is important to emphasize here that when someone begins to discuss progress, the political and cultural implications are likely to be significant and controversial.

New Technology Equals Progress:
To Question This Is Heresy

"Technological progress," a term that equates the development of new technology with progress, is a powerful term with quasi-religious undertones. It should be clear how important this concept has been in the formation of the national identities of those who live in the United States, and to a considerable degree in all Western industrialized countries. However, by now it should be abundantly clear that there are serious problems with the idea of progress, especially when equated to technological development.

The term "heresy" refers to ideas or beliefs that are held in opposition to widely held, dominant beliefs of religious or quasi-religious importance. It is a powerful term: people deemed heretics have been variously burned at the stake, excommunicated, ostracized, or vilified. It is unfortunately true that in contemporary culture, ideas that are depicted as resisting progress are dismissed with scorn, and people who propose alternatives to blind adherence to the progress narrative are vilified as standing in the way of progress. Even more extreme, as David Noble has pointed out, it is very nearly a heretical act just to question the equation of technological development with progress.[25] It is almost as if it is un-American, destructive, backwards, and dangerous to even ask: is the development of new technology necessarily progress? Perhaps this is because to do so invariably raises questions about how structures of power work, what the national and international implications of this power are, as well as how our sense of identity ties us emotionally to these same structures. But question we must.

To understand the power of the equation that new technology equals progress, there are two compelling questions that merit asking any time the progress story is aired. We end this chapter by posing these two questions: Progress for whom? And progress for what?

Progress for Whom?

Who really benefits if we believe the story of technological progress as it has been told to us? The answer to this question will vary depending on circumstances, but most often those who benefit are those who control the technologies or who make a direct profit from their use. The story provides popular support (more powerful than advertisers could ever hope to achieve) for the projects of science, technology, and industry. When the railroad was the symbol of progress, the railroad business was booming and fortunes were made. When electricity was the symbol of progress and projects like Hoover Dam and Tennessee Valley Authority were begun, power companies reaped the benefits.

Also, apart from the idea of direct benefit (power and profit), progress favors some sections of the population over others. If a computer is a mark of progress, those with the resources to own and operate the newest computers benefit. However, those who cannot afford the newest technologies are shut out, unable to benefit from or share in the vision of the good life.

We also have to keep in mind that progress for some may mean a burden for others. For example, for some it may seem like progress that so much more and new information can be processed and accessed by computers. But how does all that information get there? Low-paying, grueling data-entry work is the price some pay in order for others to progress. And what of the secretarial jobs that are lost because every boss now has his or her own capacity to compute? What work remains open for those displaced secretaries? Data entry perhaps? And what of the less-developed nations where increasingly data entry is being outsourced? For many of these countries development has come to mean producing sophisticated technologies, products, and services (such as data entry) for consumption in the developed nations. Thus, the menial, low-paying work of many people in the world supports much of the technological progress enjoyed by others. In what sense is this progress? It is thus always critical to assess who benefits from the progress narrative and who does not.

Progress for What?

It is also critical to assess the typically unexamined goals implied by the progress narrative and reassess them. To that end it is insufficient to simply return to the Jeffersonian balance of material and moral progress, or Emerson's choice between works and days. In addition, we ought to seek out other goals that enlarge the range of options from which to choose. Such goals might focus on democracy, community, sustainability, conviviality, spirituality, and so on. We will address some of these goals later in this book, but suffice it to say here that as we change our goals, technology's role in culture changes. It is possible—perhaps necessary—to devise different ways of assessing progress.

Although the progress narrative is alive and well in cultural practice and imagination—particularly in the form of the "cool" and the "neat"—progress no longer seems to be the term of choice when thoughtfully justifying technological decisions. It is still used to dismiss troublesome thoughts about technological decisions, as in "Well, that's progress," usually accompanied by a shrug and a sense of irony. But it is less likely to be used as an explicit reason for explaining technological decisions such as purchasing a new technology. For example, we aren't likely to justify the decision to purchase a cell phone by saying, "I bought a cell phone; that's progress!" The term used this way sounds more than a bit old-fashioned. Far more likely is the justification, "I bought a cell phone; it's cool!" However, augmenting "progress," and to some degree supplanting it, the term "convenience" incorporates and in some ways refines the notion of progress. It makes good contemporary sense to say, for example, "I bought a cell phone because it's really convenient." We turn then in the next chapter to the concept of convenience to explore its story and its role in technological culture.

CHAPTER TWO

Convenience

Convenience Is Another Story

THE SCENE: A BEAUTIFUL SUMMER DAY in a suburb with neatly clipped hedges and grass, lots of houses close together, and no sign of people. Focus: a house with an automobile parked in a blacktop driveway. A woman emerges from the front door, walks over to the automobile, and gets in. Quickly she backs the automobile out of the driveway, drives about ten feet to the mailbox, reaches out, gets her mail, backs up, pulls back in the driveway, gets out, and returns to her house. End of scene.

This vignette, from the cult film *The Gods Must Be Crazy*, never fails to draw laughs.[1] Why, you are meant to wonder, didn't she just walk to the mailbox? It might have taken a bit more time to walk to the mailbox and back, but it might actually have taken less! Present in the laughter is recognition. People in the audience invariably recognize the woman's acts as representing their own. They see in her actions their own habitual uses of technology. Why drive the automobile to the mailbox? The answer is simple: because it is more convenient. It keeps her from having to exert energy. It allows her to move faster. It makes covering distance, however short, faster. The automobile makes life easier, and that is what it is supposed to do. Why walk when you don't have to? Furthermore, convenience has become habit. When most people have to go somewhere they habitually choose some form of mechanical transportation: private automobiles, taxis, busses, subways, airplanes, maybe the limo if they are lucky.

Is driving to the mailbox progress? The story of progress, as we discussed in the previous chapter, offers some explanation for choosing to drive to the mailbox rather than walk. Technologies are developed to do things for you that you might otherwise have to do for yourself, and that's progress. But to raise once again the difficult question we raised in the previous chapter, does it make life better? Is life better if you can take the car to the mailbox rather than walk? Many people would argue that it's not. People in the medical professions might say that you need that walk, because life is better when you exercise properly. Environmentalists might say that you should walk, because life is better when you don't let automobiles use up precious resources and produce harmful emissions. Psychologists might say that you need that walk, because life is better when you take the time to slow down and engage the world. Community activists might say that you need that walk, because life is better when you meet and interact with neighbors. If you grant credence to just some of these arguments, driving to the mailbox cannot be explained solely in terms of progress. There is clearly more to your relationship with technology than the story of progress alone can account for. At least part of the relationship has to do with *convenience*.

The value and practice of convenience, the story of the desire for and attainment of comfort and ease, is another story that plays an important role in technological culture. In some ways the commitment to convenience contributes to the story of progress. But because convenience tells its own story, it can also undermine progress. Progress is a grand and formal story that accompanies feelings about big events, like the feelings of pride accompanying the announcement of the human genome sequence in June 2000. But convenience is a mundane story, an everyday, garden-variety warrant for decisions involving technology at its most banal. Convenience, more often than not, is the everyday motivation that justifies ongoing choices involving the role of technology in everyday life. The woman drives to the mailbox, not because it is progress to do so, but because it is convenient. The importance of this story in everyday life obliges us to take a closer look at the meaning and practice of convenience.

What Is Convenience?

Convenience, like progress, parades itself initially in fairly uncomplicated dress. The story goes like this: Technologies make life better because they make life more convenient; that is, they save time, conquer space, and create comfort. Technologies perform tasks we might otherwise have to do for ourselves. They relieve us from drudgery, labor, and physical exertion. They make it easier to go to more places faster. They minimize the everyday struggles that were commonplace for our ancestors. In all, they make life easier.

There is, however, much more to the story. Thomas F. Tierney, in *The Value of Convenience: A Genealogy of Technical Culture*, lays out a richer version of the story of convenience.[2] He argues that the desire for ease, what he calls the *value of convenience*, is integral to understanding the modern self and modern technological culture. As Tierney explains, convenience in and of itself is not undesirable. Indeed, it can be quite liberating, and it accounts for many of the improvements in the quality of life that characterize the contemporary world. However, convenience becomes a problem when the value of convenience and the desire to achieve convenience come to dominate technological culture. Far from being merely liberating, the effects of the quest for convenience have had widespread and disturbing effects.

The modern dominance of the value of convenience is related to a significant shift in the meaning of convenience. *Convenient*, before the seventeenth century, meant that something was in accordance with, in agreement with, suitable or appropriate to a given situation or circumstance. It also meant something was morally appropriate.[3] A convenience would thus have been something that was suitable. If something fit the circumstances, it was convenient. For example, serious winter clothing for those living in the north woods is a convenience. A board of just the right size, used to suit the requirements of a building project, is a convenience.

This notion of suitability differs dramatically from our contemporary notion of convenience. The contemporary meaning of convenience continues to denote a sense of suitability, but radically redefines its connotations. Now something is convenient only if it is suitable to one's personal comfort or ease. A dictionary definition indicates that agreement, harmony, and congruity are obsolete definitions. Suitability heads the list of definitions, buts its meaning shifts—modified by additional definitions—to insist on personal ease and comfort. Those definitions include:

> fitness or suitability for performing some action or fulfilling some requirement... a favorable or advantageous condition, state, or circumstance...something that provides comfort or advantage: something suited to one's material wants...an arrangement, appliance, device, material, or service conducive to personal ease or comfort...freedom from difficulty, discomfort, or trouble.[4]

Personal comfort obviously plays a crucial role in the connotations of convenience, and the meaning of comfort has shifted along with the meaning of convenience. Tierney points out that, before the fifteenth century, comfort referred to strength and support. To comfort, "meant to support, strengthen, or bolster, in either a physical or mental sense." In the fifteenth century, comfort also began to mean removing pain or physical discomfort. But by the nineteenth century, comfort came to mean "a state of physical and material well-being, with freedom from pain and trouble, and satisfaction of bodily needs."[5] To be comfortable, to experience ease and convenience, one must thus be free from pain and trouble and have all bodily needs satisfied. This is the current expectation most people have of technologies: Make us comfortable. Make life easy. Make life pain and trouble free. Meet all bodily needs. This last point, satisfying bodily needs, is crucial for Tierney. Understanding the changing nature of bodily needs is key to understanding the uniqueness of this contemporary role of technology.

Convenience and the Body: From Meeting the Demands of the Body to Overcoming the Limits of the Body

The changing meanings of convenience and comfort correspond to significant changes in the way people relate to their bodies. Tierney argues that between the time of the ancient Greeks and the present, the perception of what the body needs has changed dramatically. The ancient Greek household—made up of the

Greek male citizen, wife, children, animals, and slaves—was organized to produce what was necessary for survival. The body made certain demands—for shelter, food, clothing, water, and so on—and it was the task of the household to meet, or satisfy, those demands. Because Greek male citizens participated in the life of the *polis*—the political arena that has come down to us as characterizing Greek life—some scholars have suggested that they did not participate much in or value the life of the household. However, the evidence, according to Tierney, points to the fact that even the male citizens placed great value on performing the activities of the household and meeting the demands of the body.

Tierney contrasts this Greek value of *meeting the demands of the body* with the contemporary value of *overcoming the limits of the body*. Where the Greek body was seen as making demands, the contemporary body is seen as having limits. Where the Greek body was more or less a given with certain requirements, the contemporary body presents problems that need to be overcome. If we think of our bodies as having limits, we see them as lacking something, as having limitations, as falling short, as having problems that demand solutions. Our bodies get tired and sore, they can't be in two places at one time, they don't move very fast, they break down, they age, and ultimately they die. Clearly Greek bodies did this too, but the difference, according to Tierney, is that the Greeks viewed this as a simple *fact* of the body, whereas we view this as a *problem*. If having these limits is a problem, then we take it as our destiny to solve the problem. We do this by attempting to overcome the limits. We strive to find ways to not get tired and sore, to be in two places at one time, to move faster, to not break down, not age, and ultimately, to not die. And we strive to do this conveniently, that is, without pain or discomfort, without unnecessary exertion.

The interesting thing about limits is that once you conceive of the body as having limits to overcome, you are doomed to never be able to overcome them. Why? Because once you overcome a limit you automatically establish a new limit. Overcome the next one and you automatically establish another. A limit, like the horizon, always lurks out there before you, no matter what you accomplish. Take sports records as an example. Once people thought that no human being could run the mile in less than four minutes. That was the limit. Roger Bannister overcame that limit in 1954. Bannister's new record of 3 minutes 59.4 seconds was then broken by John Landy, also in 1954. Landy's new record of 3 minutes 58 seconds was also eventually broken. Currently, top runners regularly run the mile in less than 3 minutes 50 seconds. And whatever the present record, runners are out there still trying to overcome it. The current record is nothing more than a limit horizon taunting runners to overcome their imperfect bodies and exceed the limit. Once they do, however, the limit horizon will merely move its location a little further down the road and continue to taunt runners for their limitations. Whereas the Greeks satisfied bodily demands by careful household planning, we rely heavily on the development and use of technologies to overcome bodily limits. In the case of running faster, more advanced training technologies, new high-tech shoes, new high-tech running clothes, or new pharmaceuticals might be just the ticket to push past that limit. Records are meant to be broken. Limits are meant to be overcome. New technologies promise to overcome the receding limit horizon.

According to Tierney, the desire to overcome bodily limits has taken two forms primarily: *the desire to overcome the limits of space* and *the desire to overcome the limits of time.* The two are closely connected, though not identical. On the one hand, we have become increasingly frustrated with the limitations of our bodies to take us further than we have already been in a more convenient fashion (a limit of space). On the other hand, we have become increasingly frustrated with the limitations of our bodies to get us to all those places more quickly than we have been able to in a more convenient fashion (a limit of time).

Because we make space a problem, we continue to develop modes of transportation that originally were designed to exceed the limit of how far a person could walk or run in a day or a season. Now, however, the limit horizon demands that we develop technologies to take us beyond the limits of outer space. We routinely expect our transportation technologies to make it easier and more comfortable to take a quick weekend vacation on the other side of the continent, or the other side of the world. Business travel often requires people to be in one city in the morning, another in the afternoon, and perhaps a third by nightfall.

Because we make time a problem, we continue to develop technologies to get us to those places faster. Since time spent traveling is a bodily inconvenience and contemporary life demands that we get to places and back again in a limited amount of time, we have to be able to go and return quickly. Those quick weekends on the other side of the continent or world are only possible if we can do it in a weekend. We've got to be back to work on Monday, after all!

Another limit we must contend with is one that clearly combines the limits of time and space: the need to be physically present at a particular place at a particular time. Routinely, we expect our communication technologies to make it easier and more comfortable to stay in touch with any other person or place we can imagine, regardless of where we or they might be: the bath, the car, the swimming pool, the jungle, the mountaintop, or the space station. The challenge for new technologies is to collapse space and time so that the communicator/traveler can be everywhere at once without exertion. We have come to place a high value on being somewhere without having to go there. You can sit in the comfort of your chair and go to the Library of Congress to look up a book, or go to the AfriCam web site and check out the animals at your favorite watering hole in Africa. You can experience both, with a split screen, and thus be in three places at once: the Library of Congress, Africa, and home. By collapsing time and space in this way, technologies work toward (but never entirely succeed at) making all spaces equally and instantaneously present with complete comfort and ease.

Enter the need and desire for communication technologies to stand in as surrogates for our bodies in what has come to be known as "telepresence." Again, the limits have been dramatically reconfigured. Early communication technologies were designed to detach the message from the sender and send it over the hill, as with a smoke signal, or as far as a person could walk, as with a written message sent with a messenger. Now, however, the limit horizon requires that we develop technologies that allow us to communicate with others long distance in ways that reproduce our actual presence. Some of the research that is the farthest out there, closing in on the current limit horizon, is about linking virtual bodies anywhere at any time, thus enabling a variety of human interactions without interference from either time or space

The ultimate limit of the body is the limit of its lifetime. All living bodies, at least as we write, will die. Death is the ultimate inconvenience because there is widespread suspicion that we can do nothing that will ever allow us to overcome *that* limit. Conveniences can only band-aid our lives with ease and comfort within the limits of a lifetime of unpredictable length. The fact that this makes us pretty uncomfortable is evident in a variety of cultural venues. For example, the development and use of medical technologies is designed to prolong life. Advertisements for medicines and supplements sometimes suggest that one might live forever. In science fiction, people live forever in virtual reality. Cloning technologies are frequently talked about as if they were a means to immortality. If you can be cloned, isn't there a sense in which you can live forever? If death is the ultimate limit of the body, the ultimate technology will be the one that overcomes death. But, perhaps, this is like the four-minute mile, and once that limit is overcome, a new limit horizon will stretch out before the inhabitants of the future.

In the meantime, we develop and use technologies to extend our lives and make us as comfortable as possible. The eyeglasses some of us wear are conveniences that allow us to negotiate the terrain with far more ease than if we went strolling around without them. Laser eye surgery promises even more convenience, because we won't have to deal with the inconveniences of eyeglasses. We won't have to feel their irritating weight, remember to clean them periodically, wrestle with them as we put on a pullover sweater, or wipe off the steam when we go skiing on a cold winter's night.

Life, most of us would agree, is definitely better with all the conveniences of transportation technology, medical technology, household technology, communication technology, farming technology, industrial technology, and so on. But is that the whole story? No, we think not. Nothing, of course, is that simple; and beyond a doubt, the role of technology in our lives is not that simple.

Wants and Needs

Convenience does not in any incontrovertible way make life better. Like the old story of the blind men led up to different parts of an elephant and asked to touch it and describe it, how you describe the role of technologies of convenience in culture depends on where you stand in relation to its many parts. The part that most people fail to see relates to the changing nature of needs that accompanies the changing limit horizon of the body.

It is true that bodies have needs that absolutely must be met. Scholars in the social sciences often debate about the exact nature of basic bodily needs, but they are generally biological and include shelter, food, water, clothing, sleep, affiliation, and procreation. These are the sort of basic needs that the Greek household, according to Tierney, was organized to deliver. Surely the Greeks had wants—that is, things they desired that were not absolute necessities—but life was organized more around the needs rather than the wants. Contemporary human beings continue to have the very same biological needs, but over time, as we began to develop a sense of bodily limits, what we needed expanded to include nonbiological, culturally produced needs. Things that formerly seemed to be wants became, in fact, needs. Air travel provides a good example. At one time in history, nobody needed to travel by air. People certainly dreamed of the possibility and longed to be able to travel by air. But it was a want, not a need. It was a tantalizing limit out there waiting to be overcome. Once the limit was overcome and travel by airplane became possible, it became a luxury. In fact, for many people, air travel still seems like a luxury and their survival does not seem to be connected to it. However, in several very interesting ways, air travel has become a necessity.

Earlier we mentioned that business travelers are often in one city in the morning, in another by midday, and in yet another by night. If you travel in an airplane during the week, you will likely be seated among these same business travelers doing what they must, that is, working hard to overcome the limits of time and space by flying from city to city as required. Do they have to fly? Is flying a necessity? Certainly flying is not a necessity in any simple biological sense. But, if they want to keep their jobs, if they want to feed themselves and their families, if they want to fit into the mainstream of how things are done, they have to fly. Surely, you might protest, they could quit and take a job that does not require them to fly. This is certainly true, and there are plenty of people who choose not to take jobs because they would be required to fly. Okay, so what job do they take then? Perhaps they take a job that requires them to drive. But driving is not a biological necessity either, is it? So, if they don't take that job, what is open to them? We can play this game for a long while, tracking down ways that any job they might take can make a necessity out of something that is not a biological necessity. In the end, you might say the person has the right to choose to not work! And, again, you would be correct. But what kind of life is open to a person in this culture who chooses not to work? The point is, to be a fully functioning adult member of the culture, you are likely to have accepted as necessities various technologies and technological practices that are not biological, but are rather *cultural* necessities. They are necessities, nonetheless.

In this way, wants and luxuries become necessities. They become habits deeply entrenched in the way that culture is organized. Food is doubtless a necessity, but refrigeration is not. However, once urban and rural areas are organized as geographically distinct areas with distinct tasks, and there is no space in the city to garden, and it takes a long time to get food from the country to the city, then refrigeration becomes more like a necessity than a luxury. The necessity seems cultural rather than biological, but in the end the implications are biological as well. What happens if you can't get fresh food in a hot summer in a city without the aid of refrigeration?

It is interesting to speculate a little further about what happens when wants and luxuries become necessities, and these necessities entail overcoming the limits of space and time. In short, culture becomes organized around the project of overcoming the limits of the body. We increasingly need to expand our sense of the spaces we maneuver in; and we increasingly need to do everything faster. Again, business travel provides a pertinent example. In an increasingly global market, business must be able to move, and move quickly (virtually or bodily), if it is to keep up with trends. For an excellent example of this imperative, we suggest flipping through Bill Gates's aptly titled *Business @ the Speed of Thought: Succeeding in the Digital Economy*.[6] Almost every contemporary activity involves the need to collapse time and space by overcoming their limits. Researchers interested in eradicating viral disease must contend and compete with the speed at which diseases travel on global transportation systems. (The SARS outbreak of 2003 is an example of this.) Parents must contend and compete with the rapid-fire exposure to a nearly full array of worldly activities children encounter through television and the Internet. Employees have to contend with demands to relocate on short notice or travel long distance. Teachers must contend with pressures to offer courses online using distance-education technologies. What we want and need, and what we must respond to, increasingly relates to the value of convenience—to the desire to overcome the bodily limits of time and space—and technology is integral to the process.

When Convenience Isn't

The story we tell ourselves about convenience, the story built right into the meaning of the term, is that it makes life easier and more comfortable. We might think that some of the demands made on us, like having to travel by airplane or having to restrict children's access to the home computer, are the necessary side effects that we must accept in order to overcome bodily limits with comfort and ease. They are "the price we pay," so to speak. That's certainly a powerful story, but one that, again, sees only part of the elephant. Sometimes it makes more sense to recognize that convenience isn't always so convenient!

In a classic study of housework and household technology, one that we will return to later in this book, Ruth Schwartz Cowan looks closely at the relationship between household conveniences and the changing nature of work in the American home.[7] Her study suggests that using convenience technologies does not always mean that life is altogether easier. Modern household conveniences—washing machines, refrigerators, vacuum cleaners, dishwashers, microwaves, bread machines, and so on—certainly have been marketed as labor-saving devices, promising more leisure time and less physical exertion. Cowan concurs that these conveniences are part of an overall rise in our standard of living and that they do reduce the drudgery of particular tasks. It is, after all, physically very easy to walk over to the washing machine and throw in a load of clothes; but these technologies do not eliminate labor. In fact, as a part of a changing technological system, they contribute to an increase in woman's labor.

If you look past the idea that technology is just the physical stuff—the washing machine or the bread machine—you will see that household conveniences are part of a network of connections that tell a different story, one in which, as Cowan's book title tells us, there is actually *More Work for Mother*. Cowan describes the changing nature of household technology as a process of industrialization of the household, where the process of production and its products change. As part of this process, men were gradually eliminated from household production, as was hired household help. Eventually, as Cowan argues, the technological systems that define the household are "built on the assumption that a full-time housewife would be operating them."[8] Accompanying this shift, the standards of cleanliness and health increase, and become the sole responsibility of the housewife. Guilt, embarrassment, and insecurity drive household labor. Cowan claims that:

> the hard-pressed housewife was being told that if she failed to feed her babies special foods, to scrub behind the sink with special cleaners, to reduce the spread of infection by using paper tissues, to control mouth odor by urging everyone to gargle and body odor by urging everyone to bathe, to improve her children's schoolwork by sending them off with a good breakfast, or her daughter's "social rating" by sending her off to parties with polished white shoes—then any number of woeful events would ensue and they would all be entirely her fault.[9]

Consequently, clothes have to be washed more often, more elaborate meals have to be produced, more cleaning has to be done, and more products have to be purchased. From this perspective, the conveniences no longer look so convenient.

The current popularity of bread machines illustrates how more labor is demanded as part of the desire to better provide for the household with modern conveniences. If you want your family to eat healthful bread and to have it fresh and warm and lovingly presented, buy a bread machine! Oh yes, and then buy the right kind of flour, yeast, and the special ingredients for all the speciality breads that you will make if you really love your family. Oh yes, and make it fresh every day. That, after all, is what the machine is designed for. Oh yes, and clean the machine parts after use, and dust it when you clean the counters now cramped with other labor-saving conveniences. This convenience, like all household conveniences, is part of a technological system that makes us more comfortable in some senses. However, the network of connections that constitute this technological system do not, in the end, reduce labor and save time; instead, the network of connections is part of a shifting burden in which the demands to collapse time (you can make that bread now!) and space (you can make that bread here!) become, in a sense, an inconvenience. These contemporary demands are burdens, responsibilities, and stresses that can only be called *uncomfortable*. These burdens constitute a contemporary form of *dis-ease*.

As with household technologies, so it is with transportation technologies (remember those business travelers!), communication technologies (check that e-mail hourly, better yet, set your computer to beep whenever a message comes in!), even recycling technologies (buy that special five-gallon composter designed especially for use in cities!). Increasingly, we need technologies that perform convenient tasks, and those technologies are part of technological processes that are, in turn, part of changing labor processes that actually demand considerable exertion.

Industrial production, in the more traditional sense of factory production, plays an important role in changing labor processes in three ways. First, industry constantly retools to anticipate, produce, and market new and (now) much-needed conveniences: bread makers, yogurt makers, composters, air purifiers, computers, pocket memo pads, cell phones, and new and fancier automobiles. The survival of industry depends on the timely promotion of, and adaptability to, change.

Second, industrial production becomes organized internally around the value of convenience, with consequences for virtually all labor throughout the culture. In particular the practice of scientific management, sometimes called "Taylorism" after Frederick Winslow Taylor, began to transform the workplace in the early 1900s. Speed and efficiency are the key concepts in scientific management. Its goal, according to Taylor, is to train each individual "so that he can do (at his fastest pace and with the maximum of efficiency) the highest class of work for which his natural abilities fit him."[10] Efficiency, for scientific managers, means completing a desired task with the minimum input of energy, time, materials, and money. With this goal in mind, the results of time and motion studies of particular tasks were used to redesign production processes to maximize the output of human energy at the fastest pace sustainable. The production process itself thus became organized around the ideal of convenience, overcoming the limits of space and time with maximum comfort and minimum effort.

Third, industrial production is significantly transformed by Fordism, named after Henry Ford. Fordism utilized innovations in mechanization, combined mechanism with Taylorism, and instituted the continuous assembly line. As Tierney discusses this phenomenon, Fordism has significant implications for the value of convenience and for the consumption of conveniences.[11] The most significant implication is that, by rationalizing the pace of work, industry was able to increase production, generate capital quicker, and therefore retool quickly when necessary to respond to and capture a changing market. In other words, industry too could offer more, newer conveniences by operating more conveniently. Further, by demanding a steady and intense work pace throughout the workday, workers need to recuperate at home, rendering them more likely to rely on conveniences to get through to the next day. Overall, the changing nature of industrial work creates a ready market for the conveniences that industry is increasingly geared up to produce.

Neither the material things themselves nor some essential truth about human beings has determined that these conveniences should become needs. Rather, they are part of a changing configuration of contingent connections, which suggests that life could be otherwise, given other choices. Cowan argues, for example, that "[t]echnological systems that might have truly eliminated the labor of housewives could have been built...but such systems would have eliminated the [single-family] home as well—a result that...most Americans were consistently and insistently unwilling to accept."[12] Alternative technological systems that would have eliminated the need for the single-family home, privately owned tools, and the servitude of the housewife include commercial or communal housekeeping arrangements, kitchens, food delivery services, laundries, child care, gardens, boarding houses, and apartment hotels: all with appropriately designed and sized technologies to perform the necessary supportive tasks. Many of these—and other—alternative technological systems have been variously promoted, instituted, and largely rejected.[13] The issue of choice is not always obvious. Does giving up the single-family home, with its excess of privately owned tools, seem like a choice? It is, but because it has become a cultural habit, it doesn't seem like a choice. When cultural habits become ingrained, when media offer up versions of what life should be like, when everyday economic circumstances encourage certain choices, when peer expectations exert pressure, and when political rhetoric and political practices assume one direction and not another—the chosen path may seem like the only way to go.

The Time and Space of Consumption

The path we continue to take with fervor—the path of convenience—has had monumental implications for the nature of private and public spaces and on the role of consumption. Both Tierney's and Cowan's treatments of the changing nature of household production and its relationship to technology reveal some of these changes. The household becomes a very private space; it becomes the production site for the work of the housewife, who in turn becomes a consumer of convenience technologies to help her carry on her productive tasks. Public spaces become dedicated to performing specialized tasks that are no longer part of the household. Factory workers produce clothes, prepared foods, modes of transportation, tools, lumber, machines, and industrial household technologies, many of which are designed for private consumption. Retail operations, which are increasingly centralized, sell the goods produced by industry. Public schools educate children. Public utilities deliver power and collect garbage. Mass media deliver news and entertainment.

Another way to look at this is to see it as part of a process of moving away from a culture organized around subsistence and toward a culture organized around interlocking dependency. As part of the relations of dependence, one of our major tasks as citizens in the process is consumption. This is especially true of the household, which becomes a primary, privatized site of consumption. But lets unpack this claim.

At first blush, it seems obvious that we have moved from a subsistence economy to a market economy. However, for a very long time, people have bartered goods and services. The North American Indian peoples, often popularly thought of as living a subsistence life, had extensive trade routes throughout the continent long before the arrival of Europeans. Coastal tribes traded fish for buffalo meat. Tribes from the area now known as the Upper Peninsula of Michigan bartered copper for products from the South. Throughout the continent, ivory, bone, and medicinal plants were traded. In popular myth as well, the American colonists lived a subsistence life; but that too is overstated. Cowan, in *A Social History of American Technology*, maintains that while self-sufficiency was a highly regarded value even in colonial America, "no colonial family could have produced all that it needed for its own sustenance."[14] These observations should be taken as cautionary notes. A move away from self-sufficiency is not simply a feature of contemporary technological culture. Humans, after all, are social animals, and it is probably a rare case in any era for a lone individual to have had no contact and exchange with any other human. If, however, we envision a sliding scale rather than a simple binary distinction between subsistence culture and a trade or market culture, we can appreciate the magnitude of what has changed. Specifically, what has changed is that limit horizon. Expectations about what technologies are supposed to do for us have become increasingly more demanding, with enormous consequences for the nature and quality of cultural life. It is as though we are no longer trying to run the four-minute mile, but a three-and-a-half–minute mile.

In some of the most compelling arguments in his book, Tierney describes the changing configuration of public and private spaces and the role of consumption in a discussion of the changing nature and role of agricultural technologies in the settlement of the western United States. The settlement of the West was largely controlled by government land sales. While the acreage requirements varied, minimum plot size was quite large: 640 acres in 1789; 320 acres in 1800; and 150 acres in 1804. A settler-farmer interested in living more toward the subsistence end of the scale would probably want about five acres of good land; but if they wanted to buy land, they had to buy the larger amount. Prices varied, but in 1789 the cost was $1 plus $1 per acre, for a total of $641. This was a substantial amount of money, and very likely it was borrowed, with interest due. That meant settler-farmers had to make the land productive fast in order to repay their debt. They did this primarily by purchasing, again on loan, farm equipment designed to handle a lot of ground fast.[15] Already, at this early point in the story, farming had moved far away from the subsistence end of the scale. It is "quaint" to think of early-American farmers as living subsistence lives; but they were already debtors and major consumers of farm equipment.

The situation continued to develop away from the subsistence end of the scale. Farm equipment continued to get more specialized, bigger, and more expensive. It was designed to cover more ground faster and more comfortably, that is, more conveniently. This, in conjunction with the dependence on and cost of rail transport to get goods to market, the vagaries of the market's ups and downs, inevitable crop failures, and increasing land taxes, moved farming further from the subsistence end of the scale. It is a rare family farm in contemporary America that does not have at least one member working as a wage earner outside the home—most likely the "housewife," who, as Cowan points out in *More Work for Mother*, would still be primarily responsible for the housework. In this situation, the need for more convenience technologies increases. It makes sense in these circumstances to purchase a dishwasher, a microwave oven, and factory-made clothes. Who has time to do otherwise?

The trend has continued in the direction of transforming farmers into consumers. In fact, most farmland is now in corporate hands. Farming has become predominantly industrial; and most of those who would be farmers have become consumers of factory-farmed food. The shift in the population away from farms is staggering:

1910–1920	32 million farmers living on farms
1950	23 million
1991	4.6 million

In 1993 the United States Census Bureau announced that it would no longer count the number of people who lived on farms. Clearly for some this is "progress," but for others it is an enormous loss. As Wendell Berry argues, "Good farmers, like good musicians, must be raised to the trade." Eventually, he argues, consumers will feel and pay the price.[16]

As it goes with farming, so it goes with many of the technological skills we depend on. Few carpenters anymore know how to do more than install mass-produced factory-made units. Far fewer home and apartment dwellers know how to fix anything that goes wrong. When it comes to conveniences, the reasonable choice seems to be to toss it and consume something new. Fewer people sew their own clothes, and fabric stores are going the way of the full-service gas station. People who work at retail stores and gas stations typically know very little about the products they sell. Fewer people make their own music anymore; most depend largely on consuming mass-mediated, highly manipulated music produced in a competitive "star" system. Again, for some this is progress that brings wonders that we could not produce ourselves, and that is certainly true; it also represents an enormous loss of community interaction, skill, and talent. The individual talents that do remain have become focused on learning to become good and canny consumers of convenience. As Tierney and others have argued, the household in general is transformed from being a site of production to a site of consumption. What we do in our homes is consume rather than create.

A Perpetual State of Dissatisfaction

A perpetual state of dissatisfaction with who and what we are is a final consequence of conceiving of the body as having limits to overcome. We can never get to where we are going fast enough. We can never go everywhere there is to go. We can never be healthy enough, beautiful enough, smart enough, or rich enough. We can never own enough stuff. We can never have enough technology. And we can never be satisfied with the fact that we die. This perpetual state of dissatisfaction fuels, and is fueled by, the production and marketing of conveniences of all kinds. Technologies of beauty promise improved textures, odors, colors, sizes, and shapes of various body parts. Medical technologies not only replace aging hip joints, but reshape noses and enhance breast size. Exercise technologies promise trimmer, healthier, more beautiful bodies, without the stigma of exercise we might get through work. Educational technologies promise to make people smarter with less effort on the part of the learner. Money-generating technologies promise wealth without work. Isn't this, after all, the promise held out by playing the stock market or the lottery? Science fiction offers us fantastic images of escaping the body and the inevitability of its death. Convenience, in the extreme forms we encounter in contemporary culture, offers the ultimate quick fix that is doomed to leave us needing yet another. Our technologies are shaped in part by that desire; they hold out promise, and they inevitably, in some form or another, fail us. There is always the next limit horizon to reach for.

What the Future Holds

It is an interesting situation to be in, isn't it: to be committed to conveniences that aren't always convenient, and always striving for what is perpetually out of reach? Why, we have to wonder, do we persist in our commitment to this contradiction? It might be because yet another cultural value is slowly replacing both progress and convenience as the dominant explanatory value behind the cultural commitment to technological development. Rosalind Williams in *Retooling* argues that the "progress talk" that once dominated technological discussions has been replaced by what she calls "change talk."[17] The simple, primary value of change renders irrelevant any expectation that change is supposed to get us something: the good life as progress would have it, or ease as convenience would have it. Instead, the "change journey," a journey with no reason or end other than itself, is what matters. To change, in this view, is the point, pure and simple. To the degree that the commitment to change rearticulates both progress and convenience, we are likely to witness a culture investing heavily in technological development with rampant disregard for any ill effects in its wake.

CHAPTER THREE

Determinism

IN HIS COMEDY ROUTINES, British comedian Eddie Izzard carries on a running gag about the National Rifle Association's attacks on gun control. In response to the NRA's claim that "guns don't kill people, people do," Izzard quips, "but I think...the gun helps, you know? I think it helps.... Just standing there going 'bang!'.... That's not going to kill too many people, is it?"[1] Izzard takes the ribbing even further when he asks, what if you gave a gun to a monkey? What would happen then? The NRA would have to amend the argument to say that "guns don't kill people, people and monkeys kill people."[2] In yet another flight of Izzard antics, he points to the fact that it isn't really even guns, or people, or monkeys presumably, that kill people, but bullets ripping through flesh![3]

Izzard has a point: The gun makes a particular kind of killing possible; and it is a lot easier to kill someone with a gun than with an icy glare or even with your bare hands. But so too does the NRA have a point: Guns don't go roaming around the world on their own killing people. People use them. They pick them up, aim them, pull their triggers, and, if their aim is good or if they are just lucky (or unlucky), they kill someone. On the other hand, Izzard has yet another valid point: Guns are often involved in killings where there was no intention to kill. You have to wonder if children, like monkeys, would be considered responsible for the deaths they might cause with a gun in their hands? Also, who or what is responsible if a gun falls over, fires, and kills someone? Nobody, in this case, even pulled the trigger. With regard to guns, how do you sort out these questions: What causes what? Who or what is responsible?

It is unfortunate that people sometimes think that simple slogans, like "guns don't kill people—people do," provide answers to these complex questions. Slogans like these get used—like weapons—as though they settled everything. If you talk about gun control with someone who is opposed to it, they will often offer up the slogan, "guns don't kill people, people do," as though it ended the argument. Like magic, slogans conceal the complexity of the arguments buried deep within these serious and sometimes humorous exchanges.

In fact, the issues raised by Izzard's imagined exchange with the NRA reveal a lot about how most people understand the relationship between culture and technology. Most significantly, it reveals the degree to which *questions of causality* dominate what matters in this relationship. First, something causes (or determines) something else: Guns kill people (a pro gun-control position). Or people kill people using neutral instruments like guns (the NRA position). Or a kind of partnership between the gun and humans kills people (the Izzard position). Second, the attribution of causal power is what permits the distribution of blame or praise: Guns are to blame. Or people are to blame. Or guns and people are to blame.

Regardless of these differences in the attribution of the causal agent, and in the distribution of blame or praise, the fact remains that understanding technology in terms of such attribution and distribution is the predominant way that the relationship between culture and technology is understood. As Langdon Winner wrote in his classic work on technology, *Autonomous Technology*: "In a fundamental sense, of course, determining things is what technology is all about."[4] This is as true for guns as it is for any other technology. All technologies are widely understood as being significant in terms of the effects that they have, or in terms of being effects themselves. For example, automobiles are associated with a range of effects worthy of both praise and blame: shortening travel time, increasing mobility, causing accidents, creating pollution, and so on. Alternatively, automobiles can be seen as the effect of the expansion of the cities, the movement of populations to suburbs, and the isolation of the individual in capitalism. Televisions are associated with providing access to information, educating children, entertaining the population, encouraging violence and promiscuity, lowering standards of taste and intellect, and contributing to the isolation of the population. Alternatively, televisions can be seen as the effect of increased leisure time, the need to create a national identity, and the industrialized production of communication technologies.

In *Metaphors We Live By*, George Lakoff and Mark Johnson argue that causation, the idea that there are causes and effects, is one of those basic human concepts "most used by people to organize their cultural and physical realities."[5] This is certainly confirmed by the prevailing tendency to think of the relationship between culture and technology in terms of causality. While it is simply *not* the case that determining things is *necessarily* what technology is *all* about, conceiving of the relationship between culture and technology in causal terms plays such a powerful cultural role that it deserves careful scrutiny.

In this chapter, then, we look at the commitment our culture has made to think of and respond to technology in causal terms: to the questions of what causes what, and who or what is responsible. First, we look at the dominant variant of the causal relationship between technology and culture: that technology causes effects. This approach is sometimes called *technological determinism*. Second, we consider the flip side of that commitment: the variant that holds that culture causes technology. This approach is sometimes called *cultural determinism*, sometimes *instrumentalism* or (in a particular variant) *social constructivism*. We conclude with a critique of the limitations of thinking in causal terms as we work toward an enriched sense of culture and technology.

Technology as Cause: Technological Determinism

As stated above, thinking in terms of causation is a widespread cultural practice. So it is not surprising that thinking about technology usually invokes causal thinking. The most common form it takes is called *technological determinism*, which means that technology is understood to have effects and that those effects are the principle determinant of cultural change. It stands to reason that if you think that technology is central to an understanding of culture, as we discussed in the introduction to this book, technological change will be seen as the major determinant of cultural change. Langdon Winner explains that technological determinism is a belief that depends on two hypotheses:

(1) that belief that the technical base of a society is the fundamental condition affecting all patterns of social existence and
(2) that belief that technological change is the single most important source of change in society.[6]

The first hypothesis asserts the strongly held cultural belief that technology is central to defining what culture is. The second hypothesis asserts the strongly held cultural belief that technologies cause effects and that these effects are the primary cause of cultural change. From a technological determinist position, certain key technologies are even considered to be "revolutionary." They define culture and have the power to completely change it.

Belief in technological determinism is widely held in Western culture. For a very long time, in fact for as long as there has been recorded history, people have been thinking about technology as primarily responsible for major cultural change. As long ago as the fourth century BC, when Greece was shifting from a culture based on oral communication to a culture based on writing, Plato expressed concern that writing might cause people to lose their memories. He wrote: "If men learn this [writing technology], it will implant forgetfulness in their souls: they will cease to exercise memory because they rely on that which is written, calling things to remembrance no longer from within themselves, but by means of external marks."[7] The argument unfolds thus: When people no longer practice their memory skills, they will no longer be able to rely on their memories to make judgments about the world. Instead, they will be forced to rely on external marks (such as writing) and the arguments of others to develop judgments. This situation renders them vulnerable to the persuasive techniques (either written or spoken) of unscrupulous individuals. Plato feared that writing technology, as a form of persuasion, would change Greek culture significantly and for the worse.

Notice the construction: It (writing technology) is the cause of major cultural change. Writing technology implants forgetfulness, it makes people mentally lazy, it causes people to cease using their memories, it makes people susceptible to persuasion, and finally, it causes major shifts in the way culture is organized and in the quality of cultural life. Eric Havelock, writing in the 1980s about the introduction of the Greek alphabet during Plato's time, claims that the alphabet was revolutionary in its effects on human culture: "The Greek alphabet...impinges on the Greek scene, as a piece of explosive technology, revolutionary in its effects on human culture." The Greek alphabet, for Havelock, caused people to have a completely "new state of mind," and thus a whole new way of life.[8]

The list of technologies that have supposedly caused revolutionary change of this magnitude is almost as long as the number of technologies you can name. Here are just a few of the more obvious examples:

PRINTING PRESS: Elizabeth L. Eisenstein, in *The Printing Press as an Agent of Change*, traces the effects of printing technology. In more than 700 pages of text, she depicts the printing press as having "left no field of human enterprise untouched."[9]

INDUSTRIAL TECHNOLOGY: That the term "Industrial Revolution" is so common is testament to the fact that people have thoroughly internalized the belief that industrial technology transformed the world, forever affecting the shape, pace, and quality of life.

COMPUTERS: Currently, people claim that computer technologies are in the process of revolutionizing every aspect of culture. This revolution has produced a veritable industry in prophesying the effects of the new technologies.

It is interesting, however, that it is not just the really big technologies (writing, automobiles, industrial technology, computers, nanotechnology, biotechnology, and so on) that tend to be understood in terms of technological determinism. Highly significant cultural effects are often attributed to lesser technologies. A student in one of Jennifer's classes insisted passionately that even the toothpaste pump was revolutionary in its effects. It is as though our habits of mind have become technological determinist to such an extent that all technologies are seen as inherently world-changing.

What is important here are less the details of the specific effects new technologies are said to produce, but that the significance of these technologies tends to be understood in terms of the effects that they have. Whether the technologies in question are writing technologies, printing presses, automobiles, computers, electronic technologies, medical technologies, industrial technologies, biotechnologies, or nanotechnologies, they are understood as changing the culture in highly significant ways. The culture changes from one kind to another, pushed and prodded by changing technologies.

If you bring this discussion back to the gun, you can see that from a technological determinist position, the gun is indeed responsible for massive cultural effects. The gun introduced revolutionary new ways to kill: quickly, with minimal effort or skill, and from a safe distance. This changed the face of combat: It is more likely to be mortal combat. This changed the way that differences are settled: There is always the reasonably accessible possibility of threatening to kill. From a technological deterministic position, it is almost as though the gun does roam about in the world on its own, affecting culture in such a way that killing with the gun is inevitable. Countless times, people have told us that the important thing to know about technology is that "once you have it, you have to use it." There is "no going back," "no regressing," "no going back to the cave." People have no power to change or control things; only technology changes and controls things. If this is the case, if technological determinism is right, then guns do kill people, pure and simple.

Technological determinism is a belief that may feel true in our contemporary experience; but it is hardly fact. Technologies do not, in and of themselves, determine effects. People create and use technologies. Effects are not imposed on us by the technologies themselves. Automobiles did not drop from the sky and force people to drive them. Televisions did not simply appear and make people watch them. Microwaves do not force people to change their eating habits. Rather, technologies do require various forms of involvement or participation of people at various stages of their development and use. There may be, as Thomas Hughes argues, a feeling of "technological momentum," that is, a powerful sense of inertia when technologies are developed and deployed that shapes, guides, or even pushes the further development and use of technology.[10] The sense of technological momentum is real: Technologies, once in place, do seem to encourage the alignment of all sorts of possibilities. But this feeling of and tendency toward momentum falls far short of the belief in a hard-and-fast technological determinism.

That being said, it is important to note how often technological determinist statements are expressed in popular discourse. Think how often you hear statements such as "computers are revolutionizing culture" or "computers are changing what it means to be human" or "television is causing violence" or "genetic engineering will create a better world." Thus, despite its inadequacies, technological determinism often organizes the way people understand and act in the relationship between technology and culture.

Technology as Effect: Cultural Determinism

Cultural determinism reverses the attribution of causal agency, so that culture is the cause and technology is the effect. Although it is perhaps less evident in popular discourse than technological determinism, cultural determinism is also quite prevalent in the ways that people understand and act in the relationship between culture and technology.

Cultural determinism depends on assumptions that are almost exactly opposite of technological determinism:

(1) that the values, feelings, beliefs, and practices of the culture cause particular technologies to be developed and used;

(2) that changes in culture result in changes in technology.

According to this understanding, as culture changes, it needs and develops new technologies to accomplish its goals. The nature of the technology thus necessarily responds to and reflects the nature of the culture.

For example, from a cultural determinist understanding, the culture is clearly responsible for both the appearance of the gun and the effects of the gun. The gun is understood to have been developed because there was, and is, a need, a desire, a value that necessitates developing a technology to kill quickly and conveniently. The gun was invented and is used in response to that need and desire. The effects of the gun—that is, killing and/or violence—follow directly from that cultural need and desire. People kill people.

An effect of thinking as a cultural determinist is the displacement of responsibility totally away from the technology. Whereas from the technological determinist position, technology is totally to blame and culture is let entirely off the hook, the cultural determinist position blames culture generally and lets technology totally off the hook. In this position, then, people, not guns, kill people. The technology is almost incidental, the mere instrument of a cultural need and desire. When people believe in this position, they often argue that it wouldn't matter if you eliminated a particular technology (like the gun) because the culture would come up with an alternative to accomplish the same end. If not the gun, then some other instrument to kill conveniently.

Refuting the cultural determinist position is a little more complicated than refuting technological determinism. At the most rudimentary level, clearly, the technology can't be let off the hook entirely. As Izzard suggests, the gun "helps." It is possible to kill with the gun in ways that are unique and can't be replicated with some other technology. Killing with a gun is different than killing with a sword, slingshot, or nuclear bomb. Thus, when someone kills with a gun, the gun bears some responsibility. So, as with technological determinism, there is an important relationship between people and technology that the cultural determinist position is ill equipped to understand.

In addition, it is possible to refute the cultural determinist position by challenging the assumption that technologies, in any straightforward manner, reflect the needs and desires of the culture. As the cultural determinist position implies, the effects of technologies ought to fall completely within the range of our intentions. They do, after all, reflect needs or desires. To put it bluntly, this is all too obviously not the case. Setting aside the problem of whether or not it is even possible to identify real intentions, technologies always surprise an unprepared populace with effects that were not purported to be intended. Did anyone intend automobiles to produce greenhouse gasses, or nuclear power plants to blow up in our faces, or computer keyboards to produce carpal tunnel syndrome? How can these effects be explained from a cultural determinist position?

Indeed, a cultural determinist has difficulty explaining these problematic effects. To account for these rogue effects, people have developed complex cultural categories. Foreseen effects are called *intended effects, primary effects*, or simply *effects*. But those other effects, unforeseen and sometimes undesirable, are called *unintended effects, secondary effects, side effects*, or even *revenge effects*. Edward Tenner, in his humorously titled book, *Why Things Bite Back*, makes very fine distinctions between different kinds of unintended effects. Side effects, according to Tenner, are effects that are unrelated to the intended effects of the technology. Side effects are trade-offs. Revenge effects, which might be desirable or undesirable, are unforeseen consequences that are directly linked to the intended effects. These are not exactly trade-offs but "ironic" effects that almost always sneak in the back door with the successful implementation of the technology. He gives the example of a chemotherapy treatment for cancer. If, on the one hand, the treatment produces baldness, that is a side effect, a trade-off for a cure. If, on the other hand, the treatment causes another, lethal cancer, that is a revenge effect. Tenner breaks down revenge effects even further to capture an imaginative range of ironic effects. These include rearranging effects, repeating effects, recomplicating effects, regenerating effects, and recongesting effects.[11]

The meticulous, imaginative, dedicated effort to classify differences among intended and unintended effects directs the focus away from the decisive assumption that operates in making that initial distinction between intended and unintended effects: that the culture fundamentally, though imperfectly, gives shape to these technologies, which in turn do our bidding. It is as though the "real," "significant," or "primary" effects are the intended effects. The unintended effects are somehow less real, a sort of irritating excess of the real. This is an odd contradiction, however; for aren't those unintended effects just as real? And if they weren't intended, then the culture no longer seems to be in complete control of technologies and their effects. Thus, the proclivity to differentiate between effects and side effects tells us less about the cultural role of technologies than it does about our own cultural desire to believe in cultural determinism at the same time that we acknowledge its failure.

A final problem with cultural determinism is that it discourages any response except optimism regarding technological change, no matter the unintended effects. Indeed, as Tenner argues, "Optimists welcome [crisis] as an injection of innovatory stimulus."[12] The trick, for Tenner, is to learn to "practice the ability to recognize bad surprises early enough to do something about them."[13] Responding creatively to revenge effects stimulates further technological development, and that, if undertaken thoughtfully, can only be good, since it is a further reflection of the potential to give shape to the world. "In the long run," he concludes, revenge effects "are going to be good for us."[14] We are, in the end, only always moving ahead. Differentiating between effects and side effects thus has the power to minimize whatever is undesirable about technology by favoring and highlighting the potential for positive change.

Technological versus Cultural Determinism

It is interesting that very few people will maintain a purely technological determinist position if you can get them thinking about it at least a little bit. On a theoretical level, most people will acknowledge that in most cases somebody has to pick up and use the gun for it to do anything. If you find a gun and put it in a closet, you might keep it from doing something. You know that the gun does not have a completely independent will. You know that the NRA is in a way correct to say, "Guns don't kill people." Similarly you know, at some level, that even though the computer seems to be changing cultural life rather dramatically, there are places that it cannot touch without your participation. For example, provided that you choose to do so, you can retain spiritual beliefs that are unaffected by the computer.

Similarly, very few people will maintain a purely cultural determinist position if you can get them thinking about it at least a little bit. Doesn't the theoretical distinction between intended and unintended effects really undermine the very notion that technology is merely an instrument of cultural intentions? Technologies do seem to participate in changes in our lives, whether those changes were intended or not. It doesn't matter whether you call an effect an unintended effect, a side effect, or a revenge effect. Equally, they are all effects. Both intended and unintended effects make demands on, and reconfigure, cultural life. The gun in the hand of a child can kill unintentionally, but what difference does the distinction make? The gun certainly doesn't care if it was intentional or not; and the intentions of the one pulling the trigger don't alter the fact that the person killed is dead either way.

Further, the fact that unintended effects can only be identified in retrospect suggests that the cultural imagination and its goals are hopelessly limited. No technology can ever be purely a response to easily identifiable, straightforward cultural intentions. Technologies are not mere tools fashioned just to serve culturally acknowledged needs and goals. Nowhere is that more obvious than in the myriad examples of unintended effects.

In spite of the fact that most people would be willing to admit to these observations on a theoretical level, most people still live as though one or the other—technological determinism or cultural determinism—were true. There is a tendency to see technology as either pushing culture along or responding to our cultural will. And for the most part, people come down on the side of technological determinism. But the very forced choice between technological determinism and cultural determinism is, we think, a sort of "Hobson's choice," meaning that a person must choose between options whose difference is superficial. [15] In making the choice, you've been forced into an undesirable position. You may be forced to make a choice, whether you like it or not, but in the absence of meaningful alternatives, both choices are equally bad. For example, in the movie *Sophie's Choice*, a woman is forced to choose which of her two children will be put to death. [16] In this Hobson's choice, the superficial appearance of choice is meaningless: Either choice is equally horrible; her alternatives do not make a *real* choice possible.

If technological determinism and cultural determinism are the only choices open to you, you have no real alternatives. Both of these positions rely on a simple determinism that quickly fails to provide the nuances required by responses to real-life situations. What choice do you have if you must decide whether guns kill people or people kill people? This Hobson's choice leaves no way to understand how it is that people come to develop and use guns or how guns and people play roles in a struggle to define what it means to kill, or for that matter, what it means to own a gun. To put this very concretely, technological determinism and cultural determinism would not help you parse out responsibility in the 1999 Columbine High School massacre, where two high-school students besieged the school and killed—with guns—many of their fellow students. Is the culture the cause and therefore responsible? Is the gun the cause and therefore responsible? Neither of these choices seems entirely satisfactory. We submit that thinking so restrictively—in terms of cause and effect—is an insufficient way to understand the complex processes within which guns (or any other technology) play a cultural role. Determinism is, simply put, not a helpful way to get at the questions that matter about technological culture.

So the good news is you don't have to decide between technological determinism and cultural determinism. This is not to say, however, that you can simply vacillate between the two positions based on the argument you want to make at a particular moment. Many people do this in everyday life without acknowledging the incommensurable nature of their positions. The challenge for us is to provide you with a better way of understanding the role of technology in culture so that you no longer need to resort to the determinisms. We introduce this option in Section III of this book. Nonetheless, it is important to realize and observe how pervasive are the assertions of these two positions. Both technological and cultural determinism are prevalent in everyday discourse; and when they are, questions of who or what is in control dominate concerns about technology. In the next chapter, then, we turn to the issue of control, to highlight the workings of the widely held commitment to determinist discourses.

CHAPTER FOUR

Control

VICTOR FRANKENSTEIN HAS A PROBLEM—several problems actually. He is being shunned at school, his health is failing, his fiancée of many years wants him to come home, and then there's his work. Frankenstein has created a monster, literally, out of pieced-together corpses, and he has managed to breathe life into it. The creature, however, is not what he expected, and he has fled in horror, leaving the creature to perish. It hasn't perished. Rather, it has survived and thrived, and now it promises to wreak vengeance on its creator, to be there on Frankenstein's wedding night and destroy his family.

It's a familiar story, told again and again through films and popular culture. We often mistakenly think that Frankenstein is the name of the monster; but in this perhaps we are not far off. The Frankenstein story, written by Mary Shelley and published in 1818, has become emblematic of a particular problem: the belief that we have no control over the things we create.[1] We learn this lesson first with children, of course, who refuse to obey us. But this analogy is carried further to other creations of humankind. *Frankenstein* was not the first such story. Fables about magically conjured creatures, such as golems, stretch back into mythology. The Frankenstein story has stuck with us for almost two hundred years, partly because the creature in question is the creation of science, not magic. It is a fable about the ethics of science and the control of technology. The irony here is that modern science and technology often intend to control nature or culture. Thus, to lose control of the very things that promise control seems dire.

In this chapter we discuss both halves of this argument: how technologies are perceived as the means of controlling nature and culture; and how technologies are perceived as escaping human control. After setting out the groundwork with these two positions, we visit a particularly potent metaphor for our relationship with technology: the Master and the Slave. Through this metaphor we discuss the ideas of *technological autonomy*, *technological dependence*, and *trust*. Even in an era of new technology—of artificial intelligence, expert systems, nano-technology, bio-technology, and so on—the ghost of Frankenstein rears its head.

The popular version of the Frankenstein story conveyed by films (including James Whale's elegant films of the 1930s, the Hammer horror films of the 1950s and 1960s, Mel Brooks' comedic yet impassioned and surprisingly respectful parody, and Kenneth Branagh's torrid version) is a simple monster-on-the-loose or revenge story.[2] But Shelley's book (and Branagh's film touches on this) is more significantly about the question of humans' responsibility for their creations. After his "birth," the creature confronts Frankenstein to request information about his existence. He asks Frankenstein to show compassion and create a companion to assuage his loneliness, but Frankenstein will have none of it. The havoc that occurs is not entirely the creature's fault, but neither is it entirely the creator's fault. The lesson to be learned is that we cannot disown the things we create. Langdon Winner crystallizes the lesson of *Frankenstein* with this statement: "the invention of something powerful and novel is not enough. Thought and care must be given to its place in the sphere of human relationships."[3] Technologies, the fable teaches, are never neutral or autonomous objects. They are, instead, more like creatures themselves. Only by (incorrectly and naively) viewing technology as neutral and autonomous can the creator be let off the hook. Only, for example, if the gun is neutral and autonomous, can gun manufacturers be considered completely innocent of what people do with their products. If we consider technology to be culturally embedded, we cannot so easily wash the blood off our hands.

Yes, We Have Mastery of Our Tools

Writing in the 1960s, Marshall McLuhan argued that technologies are extensions of human faculties. The wheel is an extension of the foot, the book an extension of the eye, clothing an extension of the skin, and electric circuitry an extension of the central nervous system. Technology becomes a means—a medium, in McLuhan's phraseology—to carry out that faculty. The technologies of the world become a means of carrying out human will. However, if technologies are extensions of our perceptions and abilities, then changing our technologies changes how humans perceive and interact with the world; in many ways, this changes humans themselves.[4] Though at first blush this approach seems blatantly technologically determinist (media determine what we are), McLuhan's fundamental point is how technologies—media in particular—extend our influence on the world around us.

We typically think of technologies as being key to early human survival. Weapons helped humans kill game, and digging implements helped humans find roots to eat. Eventually tools helped humans systematize their food production: growing crops instead of finding them, and herding animals instead of hunting them. In this way technologies have given humans an advantage in the basic struggle against nature. Nature, of course, refers to more than just food sources, but also the environment. Construction of houses and buildings and the domestication of fire helped humans to shape the spaces in which they lived. Once this initial battle against nature was further under control, humans could begin to devise ways to control each other. The following sections describe how nature once again became a target of human control in the modern era and also the means of controlling society.

Control over Nature and the Environment

Early in the twentieth century, the philosopher Max Scheler pointed out that science and technology were not exempt from a will to power, and that a will to power was connected to the fundamental values of that society. In the feudal period, he said, the power-drive was focused on other people (as we shall see below), but in the modern era the power-drive is focused on nature. The domination of nature, he argued, is a fundamental value of Western culture.[5] This value is deeply embedded in the idea of progress, which we discussed in Chapter 1. This is made clear in ideas such as Manifest Destiny and in images such as that of Progress striding across the landscape bringing light, order, and technology to the wilds of nature.

The examples of the technological domination of nature are numerous. We will begin with the largest, the reshaping of the landscape, and turn to the smallest, genetic manipulation. The control of nature is no more evident than in the building of large dams. The great rivers of the world—the Nile, the Mississippi, and so on—have been brought "under control." Unpredictable floods are mainly a thing of the past, rates of flow are carefully controlled, and the paths the rivers take are carefully managed. Even one of the natural wonders of the United States, Niagara Falls (a key example of the sublime: visitors flock to it to experience its power), can be shut off like a faucet. Other examples of the technological control of nature include agricultural technologies, forestry, and mining. At the smallest level, the mapping of the human genome and the capabilities of gene splicing have opened the possibility of instigating and controlling genetic mutation, allowing one, for example, to eliminate genetically transmitted disorders. It is predicted that nano-sized robots, about the size of a few molecules, will be able to enter bodies and cure and rebuild us cell by cell.[6]

What aided this view was the *objectification of nature*. Rational, scientific methods made it seem possible to turn nature into an object of study. The task of objective science was to unlock the secrets of nature—the nature of life and death, how things work, how things are related—by systematizing information and carrying out carefully planned and recorded experiments. Scientific observation requires that we set something at a physical distance (even if it is the distance in a microscope) and a psychological distance. By observing nature and other humans in this way, they become mere objects to be manipulated and understood, and not agents in their own right. It also separates humans from nature, which, supported by Judeo-Christian religions, progress stories, and economics, facilitates the view that nature is intended for human use.[7]

When we think of nature as a resource, we participate in this view. The term "natural resources"—meaning oil, lumber, ore, and so on—is an economic one. This view is a utilitarian view. Utilitarianism focuses on the use-value of objects (and people) and asks what profit can be made from something, or how something can help society grow. We ignore things we think are useless—things that don't have a specific purpose or function for that society. In terms of nature, we think of its use-value as primarily economic. We ask: How much is it worth? Many contemporary environmental struggles are over just this view. One group looks at a forest and sees it as so much lumber (a useful object) that can be sold for a particular profit. Another group sees a forest as being a home for wildlife; they evaluate the forest through a different value system.

Typically, the rational application of scientific principles, often cited as the definition of technology, is based on the idea of the domination of nature. Technology as a rational system is a system of domination and control. This was Frankenstein's view as a scientist: He rationally figured out how to reanimate a human body. The supposed infallibility of his view—his faith in science as producer of true knowledge, his rational deductions about the nature of the being he was to create—kept him from considering the possibility that the creature might not obey, or might be something other than what he envisioned.

Social Control

The rational framework for viewing the world allows us to organize and control nature: to classify nature into genuses and species, to manipulate its growth and development. It also allows us to control each other. Historian Lewis Mumford has argued that we should think of early cultures as a type of machine:

> Now to call these collective entities machines is no idle play on words. If a machine be defined, more or less in accord with the classic definition of Franz Reuleaux, as a combination of resistant parts, each specialized in function, operating under human control, to utilize energy and to perform work, then the great labor machine was in every aspect a genuine machine: all the more because its components, though made of human bone, nerve, and muscle, were reduced to their bare mechanical elements and rigidly standardized for the performance of their limited tasks. The taskmaster's lash ensured conformity. Such machines had already been assembled if not invented by kings in the early part of the Pyramid Age, from the end of the Fourth Millennium.[8]

The coordination of populations in the accomplishment of a task (for example, building a pyramid) is an example of what Mumford would call a megamachine. The model for this kind of control was the military, where ranks of soldiers work together, like an efficient machine, toward one task. Mumford writes, "[T]hrough the army, in fact, the standard model of the megamachine was transmitted from culture to culture."[9] To aid in the function of this megamachine, each element in it (each person) was given a particular position and function. A rigid hierarchy was put in place, and each level was given different responsibilities. Units specialized in particular tasks and were trained to perform their duties efficiently.

The connection of the military to control is much more than a historical aside. Technologies of destruction allow leaders to intimidate and threaten populations into submission. State organizations like the military and police take advantage of these technologies for maintaining control. It is not a coincidence that great technological strides are often made during times of war. Standardized production, the practice of triaging patients in medical care, and the development of penicillin are all indebted to war.

Less corporeal means of controlling the population were developed toward the end of the eighteenth century, when control was established through the means of surveillance. We often think of surveillance as simply watching someone, which in itself can be an effective means of control. But it can also refer to the gathering of information on people through means other than direct observation.

In order to better control their workers, who often worked on their own time in their own homes, capitalists created the factory, which brought all the workers under one roof. In this way the work process could be controlled. Work could be observed and regulated. Distractions could be minimized and workers could be expected to put in their time. The ultimate expression of this kind of control is the panopticon designed by Jeremy Bentham. Inspired by the plans of a relative's new workshop, Bentham created what he felt would be the perfect machine of social control. He designed a unique prison, which he called the *panopticon* (meaning *all seeing*). The prison was designed as a semicircle with the cells lining the walls. In the middle of the building was a central guard tower. The interior of each cell was readily observable from the central guard tower; and by means of reflectors and lights, each cell could be immediately illuminated. At the same time it was impossible for the cells' occupants to see into the central tower. The prisoners knew that they could be watched at any moment of the day or night, but they could never be sure when they were being observed. The threat of inspection rather than the threat of direct violence was thus the means of control. The threat of constant inspection meant that prisoners would have to behave correctly and that these behaviors would have to become habit. The prisoners would internalize the control and discipline themselves.

Bentham believed that the idea of the panopticon applied beyond the walls of a prison. He believed his machine would ensure social control in workshops, schools, and virtually every other institution or setting. He even devised plans for a series of panoptic villages. French philosopher Michel Foucault, commenting on Bentham's invention, writes, "Whenever one is dealing with a multiplicity of individuals on whom a task or a particular form of behaviour must be imposed, the panoptic schema may be used."[10]

Sociologist Max Weber described a further development in the technologies of social control: bureaucracy.[11] There are two primary elements of bureaucracy: the rational organization of an institution and the collection of information. Both involve technology in significant ways. Like the rational organization of the military, bureaucracies strive to organize their workforces according to the principles of rationality and efficiency. Rigid hierarchies are maintained, and work proceeds in an ordered manner. Each employee has a particular task or set of tasks. A particularly potent variation on rational bureaucracy is Frederick Taylor's notion of scientific management, introduced in Chapters 1 and 2, which focused on the organization and division of labor and the observation and training of laborers.

Scientific management, or *Taylorism*, is, in short, an attempt by management to control what workers do. One of its fundamental principles is the removal of decision-making abilities from the shop floor. Only managers make decisions; laborers only carry out their orders. The reasoning is this: A worker will only work at maximum efficiency if constantly observed and if not interrupted by the need to make decisions.

By removing decision-making powers, management engages in what is called the *deskilling* of the workforce. A knowledgeable, decision-making skilled worker is never fully under management's control. Therefore it is in management's interests to learn the worker's skills, train others in those skills, or, better yet, create a machine to replicate those skills. The most dramatic examples of deskilling workers as a means of controlling the workforce involve the introduction of machinery in the workplace, especially more modern introductions of computer-driven robotic machines. Langdon Winner tells the story of the McCormick Reaper Manufacturing Plant, which installed expensive manufacturing machines on the shop floor so it could fire key workers and break the influence of the worker's union. Once the union was destroyed and management regained control over the workers, the machines were removed, because they were too expensive to run and produced a product inferior to what the workers produced. Although the cost was great, gaining control must have been considered worth it.[12]

What is collectively referred to as "paperwork" is another significant aspect of bureaucratic control, and entails principles of rationality and efficiency in the collection of information. Paperwork refers to the records and information collected by an organization, which is designed to make it function more efficiently. Information, whether gathered through panoptic inspection or through the careful accrual of bureaucratic dossiers, must be collected, stored, and made (selectively) accessible if it is to serve a control function. This means that information technologies—including filing cabinets, recording devices, and the computer—are in another way the tools of social control.

As extensions of our human faculties and as tools of social control used in the interest of surveillance and bureaucracy, technologies seem to do our bidding. They seem, for better or for worse, to give us control over nature and society. Yes, it seems that we have mastery of our tools.

No, Our Tools Are Out of Control

For almost every example of how technologies have allowed humans to gain control of nature and society, we can think of counter-examples where technologies seem to have moved beyond the control of individuals, sometimes creating disastrous unintended consequences. Whenever we've thought we understood nature, nature comes roaring back. For example, all our dams and flood-control technologies have not eliminated disastrous flooding, as the occasional, disastrous flooding of the Mississippi river illustrates. Indeed, often flood-control measures—once they fail—exacerbate floods. Also, whenever we feel that we have established sufficient social control, people rebel. Late in 2000, for example, a bill was scheduled to go before Congress that would restrict conditions that necessitate debilitating repetitive motion in the workplace. Finally, our tools themselves sometimes seem to have lives of their own, suggesting that they are out of control. Who among us has not at some point complained about our computers giving us a hard time?

Both perceptions—that technology is firmly in our control and that it is slipping out of our control—are widespread in our culture. We may even feel both ways at the same time, or feel differently in different contexts. Speaking metaphorically, when we feel in control we sometimes say that we are "in the driver's seat," and mean that the machine is under our control. To continue the metaphor, however, don't we occasionally get the feeling that though we are in the driver's seat, none of the pedals seem to work very well (the brakes are soft, the steering is loose) and the car seems to be driving itself? At other times don't we feel like our cars are out to get us?

One way to understand this is by utilizing Mumford's idea of *megatechnics*, the notion that society can be viewed as a well-integrated megamachine. Recall from our earlier discussion of megatechnics that society as megamachine is a means of controlling a population; however, like the military, it is not a democratic means. The individual subjects who work in the megamachine, who carry out its specific tasks, and play its specific roles, don't always have a say in what those tasks or roles entail. For the majority of the population, the megamachine is a way of being controlled, not of controlling. Technology is in someone else's hands. Often it seems as if the system is running itself. Just as with modern bureaucracies, we cannot always identify the individuals on whose shoulders decision-making lies. The decisions are the result of the system itself, and it is difficult to argue with a system.

We've shifted language here to use the more recent term "system" to describe both Mumford's megamachine and modern bureaucracy. A system is a complex organization composed of interrelated, interdependent parts. As systems become more complex and more parts are added, it becomes harder to keep track of, and therefore keep control of, the work that it does. For example, as a corporation gets larger and adds employees and divisions, it becomes more difficult to keep track of who is doing what and how all the parts are connected. To use the example of a car engine, as more parts are added—fuel injectors, computerized monitoring, catalytic converters, and so on—the engine becomes more complex, and it becomes more difficult to keep track of what all the pieces are for and how they interact. If something happens to one part of a system, others parts are frequently affected; but it is often difficult to predict or track those effects. In a very complex system, it is often impossible to predict what effects a small change might have throughout the system.[13] Complex technologies—including missile defense systems, computer systems, and bureaucratic structures—function beyond the immediate knowledge and control of, except perhaps for a few experts, any one person. If the experts, commonly called technocrats, are the only ones who understand the system, there is less opportunity and less willingness for others to influence decisions made regarding those technologies. This makes the system still more authoritarian and even less democratic.

As the system becomes more complex, new technologies have to be invented to control the megamachine. With the Industrial Revolution and the harnessing of steam power, machines literally began moving beyond human control. They were stronger and faster, and capable of increasing destruction if control was lost. The railway offers an illuminating example. The steam engine could propel a train faster than the fastest horses and for a longer period of time. So amazing was its power, it was considered the symbol of progress, as we discussed in Chapter 1. But once at full speed, a steam engine was almost impossible to catch up with to warn it of impending collisions. This situation created what has been called a "crisis of control," where control over the technology seems lost.[14] To win back control, a faster technology was needed to help coordinate and communicate with the trains, or at least with stations ahead of the train. Around this time, the development of the telegraph (originally a military invention for coordinating multiple distant armies) served this purpose. Other technologies of accounting were needed simply to keep track of where all the trains were at a given time, because the plethora of trains, tracks, and schedules contributed to the crisis of control. Historian James Beniger cites examples of perfectly good train cars sitting idle for months at a time because they had been lost by the system.[15]

Modern technologies of management, communication, and information processing have become crucial in solving (at least to some extent) the ongoing crisis of control. With the recent boom in the Internet, World Wide Web, and information technologies, we are threatened with being swamped with more information than we can possibly process or judge. This too creates a crisis of control. To win back control, new information-filtering technologies, such as intelligent-agent software, are created to sort this information and give us just what we think we need. Again, these technologies are meant to solve (at least to some extent) this crisis of control.

Another way to understand our sense that technologies are no longer in our control is to focus on what Edward Tenner has called "the revenge of unintended consequences," which occurs when technologies cause more problems than they solve, or when they solve the problem they were meant to solve but create new ones.[16] For example, as we discussed in Chapter 2, the results of Ruth Schwarz Cowan's research demonstrate that domestic technologies—designed to save labor in the household—actually increased the amount of time women spent on housework.[17] Other examples raised by Tenner (his book is filled with fascinating examples) include the so-called paperless office,[18] the idea that with the introduction of networked computers, all documents—memos, letters, forms, and so on—would be electronic and distributed electronically. There would be no need for the great piles of forms and papers that accumulate in the traditional office. However, Tenner points out that offices that have become computerized use more paper, not less. Why? Because computers and copying technologies have made it easier to produce multiple copies and multiple versions of paperwork, and the reconfigured systems demand their production. Another example involves the intent to make work more convenient by telecommuting. Computers and the Internet make it possible for workers to work at home by dialing in to the office. Because there is no commuting involved and the worker is allowed to work at home (or elsewhere), the worker can manage time and resources better, work without direct supervision in relative comfort, and regulate their work schedule accordingly. However, research suggests that telecommuters end up spending significantly more time on work-related tasks than do people who go to the office. Rather than being a convenience (see Chapter 2) the new technologies make it easier—sometimes imperative—to continue to work on evenings and weekends. One final example of unintended consequences is the story of kudzu, a Japanese vine that grows rapidly and is excellent for shoring up poor soil. The US Army Corps of Engineers thought that this plant would help greatly with a soil-erosion problem in the southern United States, especially along roadsides where the clay soil washes away; so they planted kudzu across the South. The problem is, kudzu has no natural predators in the United States; and because of its rapid growth and heartiness (the qualities for which it was chosen), it has overtaken millions of acres of woods and fields. It is tenacious and very hard to kill: nature's revenge!

Just as the previous section highlighted the argument that we do have control over our technologies, here we have highlighted the opposite: that our technologies have control over us. There is no simple resolution to the conundrum of control, no way to decide once and for all which is true, because to do so would depend on the misguided belief that technology and culture can be separated from one another and that one or the other can exert complete domination over the other. Rather, as the metaphor of Master and Slave illustrates, the attempt to assign the status of dominant (autonomous) Master or subservient (dependent) Slave to either technology or culture, while a wide-spread and powerful cultural habit, is, in the end, futile.

Master and Slave: Trust and the Machine

Autonomy

When you have a complex system that uses machines to control machines, the human is "once removed" (sometimes several-steps removed) from direct control of a technology. For workers in the factories of the nineteenth century especially, the big machines seemed well out of their control. Workers often felt helpless in the face of those machines. In the terms of Karl Marx, these workers were alienated from the means of production:

> An organized system of machines, to which motion is communicated by the transmitting mechanism from a central automaton, is the most developed form of production machinery. Here we have, in the place of the isolated machine, a mechanical monster whose body fills whole factory floors, and whose demon power, at first veiled under the slow and measured motions of his giant limbs, at length breaks out into the fast and furious whirl of his countless working organs.[19]

It almost seems as if technology here has become autonomous: that it moves on its own, develops on its own, and controls itself. It is not only Marx who had this view. The idea of autonomous technology has a long history in the West. Langdon Winner, who has traced this history, argues that a sense of technological determinism, plus the sense that technology is out of control, have played a prominent role in modern political thought.[20] The issue of control has shifted. Where once we felt that we were masters of our machines—we made them to work for us, machines were slaves—the continuing crisis of control makes it seem as though it is we who have become enslaved. We have become far too dependent on our machines.

Dependence

The idea of technological dependence is fairly simple: It is the belief that we rely on technologies in so many aspects of our lives that we cannot function or even survive without them. A fairly clear statement of technological dependence was made by Theodore Kaczynski, the Unabomber (more on him in Chapter 7), who wrote: "What we do suggest is that the human race might easily permit itself to drift into a position of such dependence on the machines that it would have no practical choice but to accept all of the machines' decisions."[21]

The panic around the so-called Y2K (or Millennium) Computer Bug is an example of this drift into dependence. The Bug was a software glitch produced because old software programs only recorded the date using the last two digits of the year (1987 became 87), which worked fine as long as the first two digits remained constant. With the turn of the last century, the old software programs could not distinguish between 2000 and 1900. Prior to the turn of the century, people were concerned that this glitch would cause errors, crashes, and even destruction: failed nuclear power stations, accidentally launched missiles, disappearing bank records, and so on. To combat the problem, almost every computer and software program had to be checked for the fault and then corrected. Some of these programs and computers were decades old, and the last programmers who understood them had long since moved on or died. In addition, the complexity of these programs often foiled attempts to fix them. Despite our assumptions about the logic and organization of engineering, modern software programs are written as millions of lines of code that are not always well organized. These programs are so complex that no one person understands how the whole program functions, or the consequences of changes made in one part on the rest (a classic problem of a complex system). The result was that one could not quickly put one's finger on the "date" section of the program to fix it.

There was a great deal of press attention to the problem in the years leading up to 2000, and considerable worry on the part of the population. Some even went so far as to purchase survival gear, guns, food, gas generators, and so forth. Part of the general cultural anxiety about the Y2K bug entailed a realization, on the part of the population, of just how much their lives depend on technologies, how many of these technologies have computers in them, and how far out of their control those technologies are. Suddenly, people were worried that their VCRs, coffee machines, bankcards, and telephones wouldn't work anymore, that their everyday lives would at least be disrupted and perhaps collapse.

The arguments about our technological dependence stem from just this sort of realization: that we had become dependent on the technologies we thought were created to serve us, and that this fact could prove dangerous if not fatal to us. One bumper sticker during that era put it, "I'd never survive in the wild." People were asking themselves, if the power goes out, can I survive? Pushed just a bit further, what seem like questions for philosophers or science-fiction artists become of paramount importance: As machines become more sophisticated and replace human workers in more and more capacities, could machines eventually replace the entire human race? This is exactly the question posed by Bill Joy, shortly after we survived Y2K, in an article titled "Why the Future Doesn't Need Us."[22]

The flip side of dependence is trust. The worries over Y2K make us question the trust that we have placed in our machines and in the megamachine in which we live. We realize just how much we trust bureaucracies, large organizations, and complex technologies. Sociologist Anthony Giddens argues that trust in abstract systems is characteristic of the experience of being modern.[23] In modern industrial societies, we are obliged to trust in these systems. He labels these systems "abstract" because most of the workings of these systems are outside our immediate knowledge. For example, if I withdraw money from an Automated Teller Machine (ATM), I have to trust that the machine is functioning properly, that it is connected to the proper networks, and that the other networks to which it is connected will maintain my account properly. With modern electronic banking, I am no longer sure where my money actually is or even where the bank is. Is there a bank somewhere, or just a network of people and machines performing tasks? Money itself, according to Giddens, has become an abstract system. We trust that these colored pieces of paper and stamped metal have value. But they only have value if the megamachine continues to process them as we trust it will. Giddens says that we trust, not because we lack power, but because we lack sufficient knowledge of the system.[24] Trust is not the same thing as faith that the system will work, but a degree of confidence in that faith. We must always remember, however, that trust is related to risk, whether we are conscious or unconscious of that risk. In short, we still engage in risk when we trust.

Master and Slave

To understand technological dependence more fully, we need to understand the idea of absolute mastery, the idea that one can have complete control over others, including nature, technologies, and people. The slave is the figure with absolutely no control; it is completely at the will of the other. In *The Phenomenology of Mind*, philosopher G.W.F. Hegel tells a story about masters and slaves.[25] It goes like this: The human condition is marked by the struggle of person against person to achieve dominance and control. The winner of this struggle becomes the Master; the loser is either killed or enslaved. The point of this struggle is not dominance for dominance's sake, but to achieve the recognition that one is dominant. For the Master to achieve absolute mastery, it is not enough to have a Slave dependent on him; the Master also demands recognition of his position. But the Slave, being so utterly defeated, is not considered worthy enough to provide this recognition. Indeed, Slaves are not usually considered human by their Masters. In addition, by having the Slave do all the work for him, the Master becomes dependent on that Slave. It is the Slave who understands how to work and what it means to work with material reality: the earth and tools. In working with material reality, the Slave comes to a true self-understanding of who he or she is in the world, something that the Master can never do. The quest for absolute mastery is therefore self-defeating, since the Master is now dependent on the Slave and lacks the Slave's knowledge of the world and self-identity. Unlike the Master, who is not self-reflexive, the Slave realizes that we shall all die some day.

Karl Marx read into Hegel's story of the Master and the Slave support for his notion that the proletariat, the slave-like working class, would one day not only achieve enlightenment (something that their bourgeois masters cannot achieve), but also would revolt against their masters. If the bourgeoisie are so dependent on the proletariat, where does the true power in society lie?[26]

This same metaphor of Master and Slave has been applied to technology. From the very first stories about living machines—either conjured creatures such as golems or artificial humans such as robots—the issue of whether or not these creatures would turn against us has been raised. We see this in Shelley's *Frankenstein*, in Karl Capek's play *RUR* (which coined the term "robot"), and throughout the 1900s in short stories and films.[27] For example, the *Terminator* films are based on the premise that a sentient defense computer figures that the greatest danger to it is humanity in general, begins a war against humans, and creates killer robots to exterminate the remains of human resistance.[28] More recently, the film *The Matrix* is based on a similar premise: Our networked computers achieve a form of intelligence, struggle with humans for control of the planet, and enslave humans. In fact, humans literally serve the machines by becoming the batteries that power the computers. Humans thus play a completely passive and dependent role. They are slaves to the technology.[29]

AI, Expert Systems, and Intelligent Agents

Artificially intelligent computers (known as AI) may seem to be far off in the future, but those building their precursors still worry about issues of dependence and control—Master and Slave. For example, reflecting on advances in genetics, nanotechnology, and robotics, Bill Joy, cofounder and chief scientist of Sun Microsystems, asks the following question: "As Thoreau said, 'We do not ride on the railroad; it rides upon us'; and this is what we must fight, in our time. The question is, indeed, Which is to be master? Will we survive our technologies?"[30] For Joy, one of the central issues about these new technologies is that they are potentially self-replicating, and rapidly so. It is not only that we may make a machine more powerful than ourselves, or with the potential to undermine us, but that such a machine can easily and rapidly create more machines like itself or perhaps better than itself.

Another example of where such anxiety can be found is in the work on intelligent agents. Intelligent agents are pieces of software that work on your behalf. They have the capability to learn your wants and needs in order to function on your behalf on the Internet. For example, an intelligent agent could be authorized by you to seek out particular types of information, to purchase particular products, or to negotiate a business deal for you. The agent acts like a virtual butler or lackey. Though this software is not yet very sophisticated, its relative autonomy raises questions. MIT professor William J. Mitchell writes:

> Even if our agents turn out to be very smart, and always perform impeccably, will we ever fully trust them? And how will we deal with the old paradox of the slave? We will want our agents to be as smart as possible in order to do our bidding most effectively, but the more intelligent they are, the more we will have to worry about losing control and the agents taking over.[31]

Marvin Minsky, cofounder of the Artificial Intelligence laboratory at MIT, writes:

> There's the old paradox of having a very smart slave. If you keep the slave from learning too much, you are limiting its usefulness. But, if you help it to become smarter than you are, then you may not be able to trust it not to make better plans for itself that it does for you.[32]

The above positions seem to assume that intelligence is the same thing as self-interest, and therefore, that an intelligent machine will care more for its own interests than for others'. If that assumption is incorrect, the fears expressed may be overblown. Regardless, the cultural concern over the question of trust remains paramount: Can we trust our machines? And when we consider matters of trust, we do not have to venture into science fiction, with its killer robots, to touch highly significant cultural concerns. Matters of trust enter at a very mundane level: Will this machine work? Will it do what it is supposed to do? Can I trust that the bank computer will remember the deposit I made and not lose it?

Conclusion

The question of control highlights the fundamental circularity of many of the arguments about technology and its relation to humans. We create technologies to establish control, but then get upset that we are controlled by technologies. Technologies become a convenient scapegoat for problems that we have created. For example, violence on television is blamed for increased violence in society, especially among youth. But why, we might ask, is the violence there to begin with? Who decided to write about, record, and air violent acts? And why? What do those decisions have to do with the culture of violence in which we live? In asking these questions, we are not denying that television has effects on its audience. Rather, we point to the variety of other causes of violence in society alongside of and with television: a troubled economy, lack of funds for schools, a decline in the social role of religion, shifting parental styles, the availability of weapons, and so on.

If technology is conceived as a matter of control and dependence, of Master and Slave, it is set apart from human culture, treated as autonomous, then either blamed or praised. Either we have control over technology, or it has control over us; the effects in either case can be conceived as either worthy of praise or blame. Those are the only options. Either way we look at it, technology is considered as something apart from human culture. The question of control or determinism simply shifts weight and focus from one side to the other and back again. In the end, neither formulation of this relationship gets us very far in reflecting on culture and technology in ways that suggest new directions and new answers. Neither formulation provides an adequate map for understanding the complex web of corresponding, noncorresponding, and contradictory forces within which technologies emerge, develop, and have effects. It is time to shift our focus away from issues of control, dependence, and trust (as well as from causality, progress, and convenience), to think about technology in new ways, to pose new questions, and find, perhaps, new answers.

We begin this shift by first reviewing major critical positions that have developed in response to the positions, values, contradictions, and challenges that surround the discourses and practices of technology as we have described them in this first section of our argument. These critical responses are Luddism, Appropriate Technology, and the Unabomber. Then, in the final section of this book, we lay out a cultural studies approach that moves beyond these critical responses.

SECTION II

Critical Responses to the Received View

Convent of the Visitation, Water Wheel, Spring Hill Avenue, Mobile, Mobile County, AL
Photograph by E. W. Russell, 1937
Library of Congress
Historic American Buildings Survey

CHAPTER FIVE

Luddism

IN SECTION I, WE INTRODUCED WHAT WE CALL the received view of technological culture: the beliefs, practices, and experiences that constitute the dominant cultural sense of culture and technology. It is the commonsense version that most of us have been exposed to, within which we negotiate a relationship with technology. That commonsense version, we have argued, posits technology as the source of inevitable progress, as the vehicle for making life better by making it more convenient, as the driving causative force of "civilized" Western culture, and as the mechanism for exercising control in and over the world. Even those who critique technology often launch their theories from within the commonsense version of the story. In such cases, the "problem" concerning technology is the fear that technology controls us, rather than the other way around, or that progress has undesirable "side effects" that we have to deal with. However, in the received view, these problems are seen as playing the role of minor nuisance in an overall endorsement of the storyline.

We have offered criticisms of the received view as we introduced it, and have begun to introduce our theoretical alternative to it; but we have not yet laid out for you the components of our proposed alternative, which we do in Section III. Here, in Section II, we take you through what we think of as an *intermezzo*: in musical terms, a short movement between the major sections of a composition. This movement is meant to acknowledge that historically there have been important critical responses to the received view that have *not* been argued from within its logic. While there certainly have been more than the three responses we consider here—Luddism, Appropriate Technology, and the Unabomber—we have chosen these three because they cover a range of responses from which there is something significant to learn. Each is problematic in its own way; but each also offers important insight: first, into the ways people have been blinded and/or blind-sided by the received view; and second, into some of the crucial components with which we construct our approach. Therefore, even if we do not identify with Luddites, Alternative Technologists, or Unabombers (indeed, least of all Unabombers), there is something that each of these responses can offer in piecing together a cultural studies approach to technological culture.

To be labeled a Luddite, in common parlance, is to be accused of being rabidly and ignorantly antitechnology and antiprogress. Luddites, popular usage suggests, are machine haters, sometimes machine breakers, sometimes anarchists, but always dangerously misguided souls who would reverse the flow of progress and have us "go back to the cave." For example, environmental activists opposing development projects are often called Luddites, implying that they are just simply and indiscriminately antitechnology, antidevelopment, antiprogress, and therefore, anti-the-good-life. If permitted their way, the story goes, they would destroy all the good that industrial progress has brought, and render life, once again, mean, lean, and inhumane. Luddites would bring back the days of high rates of infant mortality, a short life expectancy, hard physical labor, debilitating pain, and suffering. While the efforts of Luddites may sometimes seem good natured or even quaint, they are, most people conclude, fundamentally misguided.

This characterization of Luddism as a technophobic response to new technology—and, therefore, to progress—is unfortunate, but it is hardly surprising. Given the power of the received view to frame any criticism of technology as irrational, futile, and fatuous, it makes a type of perverse sense that what is really a fascinating and instructive moment in the history of technological culture would be reduced and misunderstood in this way. An understanding of the Luddite movement, achieved by listening seriously to the issues it raised, rocks the received view to its core.

To learn from the Luddites, we turn to the careful work of historians who have been willing to look past the summary dismissals of the Luddites, which were shaped by a blind commitment to the received view. To look with fresh eyes at their history, we draw, most notably, on the work of E.P. Thompson, in his monumental study *The Making of the English Working Class*, and Eric Hobsbawm, in his meticulously researched article "The Machine Breakers."[1]

Historical Luddism

It is difficult to characterize the Luddites and the Luddite movement for several reasons. Foremost among them is the fact that it was dangerous—even illegal—to be a Luddite. During the height of the movement, Luddites were hanged. By necessity they were secretive about their activities. Second, there are no surviving, comprehensive, and written accounts by those who considered themselves Luddites, if indeed any were ever written. A few reminiscences written in the late 1800s claim to penned by or based on the stories of Luddites; but even if true, these accounts were constructed nearly sixty years after the fact.[2] The histories of the Luddites on which we draw are the result of painstaking archival research sifting though letters, press coverage, public documents, and even literature written during the period. Third, evidence suggests that the Luddite movement might have consisted of different, perhaps even relatively autonomous movements, rather than a single movement with a single coherent story. Finally, the story of the Luddites was from its inception caught up in a difficult political moment in which an allegiance to the received view of technology and culture was already at stake. Interpretations of their story have always depended on where one stood politically with respect to that view. Consequently, accounts of historical Luddism that presume to dismiss them out of hand, or oversimplify their significance, should be held in suspicion.

Luddism refers to a movement or movements of skilled workers and artisans in England in 1811–1817 in the textile industry, principally croppers, stockingers, and weavers.[3] The difficult political moment within which Luddism arose as a response was a major shift in the nature of capitalism, the changing role of workers in the development of industrialism, and the development of new technology. Prior to this time, there was an understanding that the relationship between an industry and its workers was one of mutual support and obligation. Industry provided a livelihood for its workers; workers provided skill with dedication to the craft.[4] Textile manufacturing was craft work, carried out by skilled laborers brought up through an apprentice system and protected by what Thompson calls "paternalistic legislation."[5] To be a craft worker meant that the workers themselves largely shaped the knowledge, execution, and control of the labor process. Craft work may be difficult, but it is nonetheless creative.

A crisis in this situation was provoked by the gradual encroachment of the practice of laissez-faire capitalism, which shifts the idea of mutual support and obligation by arguing that, theoretically anyway, the overall economic situation of the country improves when the owners of industry are permitted free rein to maximize their profits, and when the quality of life and work of the individual worker is not given highest priority. It is not possible, however, to discount the motive of simple greed, which government policies and cultural practice had previously curbed. Nor is it possible to discount the motive of survival in what might have been, in effect, a coercive situation. As some manufacturers developed a competitive advantage using modern factory techniques, others might have felt "forced" to do so to survive.[6] Whatever the mix of motives, the paternal relationship with workers and their independence as craft workers was seen as a hindrance to the maximization of profits. In response, manufacturers fought— eventually with success—government intervention and sought to rationalize the production process to minimize their expenses. To that end, it was desirable to exert control over the labor process by developing a factory system, replacing workers with machines wherever possible, deskilling the nature of the work, and keeping the cost of labor low.

The success of the manufacturers was hard won, and depended, in the end, on the voice and force of government adopting the voice and interests of the manufacturers.[7] It has been estimated that there were 12,000 troops deployed against the Luddites in the six counties where they were active,[8] and a number of Luddites were killed. Laws were eventually passed that resulted in deportation, jailing, and even hanging of many of them.[9] The Luddites did not set out to kill anyone or to destroy property indiscriminately; their actions had, for the most part, all the marks of a defensive rather than offensive strategy. So it is astonishing when you think about the fact that machine-breaking became a capital offense. It indicates just how strongly the culture of the time was threatened by the challenge to the narrative of progress.

But what did the Luddites do? Although it is debatable just how well organized they were, they resisted the changes being imposed on them by the manufacturers. Thompson calls them a *"quasi-insurrectionary movement*, which continually trembled on the edge of ulterior revolutionary objectives."[10] They objected to the deskilling of their jobs, the replacement of workers by machines, the extraction of exorbitant rents on the machines they used, the reduction of wages, and their overall subjection to the modern factory system in which they were treated more like servants than craft workers. Their resistance took many forms: negotiating, bargaining, striking, burning, rioting, and machine-breaking. These last are what live in the popular memory as the legacy of the Luddites: riot and the destruction of machines. But in a very real sense, their insurrectionary resistance was part of a long tradition of "collective bargaining by riot" in which rioters would do whatever they deemed effective in their effort to gain concessions, including wrecking private property, finished goods, and machines.[11] However, even though the motives of rioters would surely have been mixed, Luddite activities where characterized by legitimate motives that were widely shared. As Thompson writes:

> What was at issue was the "freedom" of the capitalist to destroy the customs of the trade, whether by new machinery, by the factory-system, or by unrestricted competition, beating-down wages, undercutting his rivals, and undermining standards of craftsmanship. We are so accustomed to the notion that it was both inevitable and "progressive" that trade should have been freed in the early-19th century from "restrictive practices", that it requires an effort of imagination to understand that the "free" factory-owner or large hosier or cotton-manufacturer, who built his fortune by these means, was regarded not only with jealousy but as a man engaging in *immoral* and *illegal* practices.[12]

Luddism was thus a highly significant "transitional" conflict, one that "looked backward to old customs and paternalist legislation which could never be revived." At the same time, "it tried to revive ancient rights in order to establish new precedents."[13] Luddites were fighting for a way of life in a changing world, and they recognized that machines, and their incorporation into a system of work, were a crucial component of that way of life.

It is perhaps a prejudice of twenty-first-century Americans to think that industrial workers in the early 1800s were probably pretty slow witted. But the history of the Luddites suggests otherwise. As Thompson concluded:

> the character of Luddism was not that of a blind protest, or of a food riot.... Nor will it do to describe Luddism as a form of "primitive" trade unionism.... [T]he men who organized, sheltered, or condoned Luddism were far from primitive. They were shrewd and humorous; next to the London artisans, some of them were amongst the most articulate of the "industrious classes". A few had read

Adam Smith, more had made some study of trade union law. Croppers, stock-ingers, and weavers were capable of managing a complex organization; under-taking its finances and correspondence; sending delegates as far as Ireland or maintaining regular communication with the West Country. All of them had had dealings, through their representatives, with Parliament; while duly-apprenticed stockingers in Nottingham were burgesses and electors.[14]

Luddites did destroy machines, but for the most part only those machines that embodied the offenses of the way of life they saw being forced on them. In case af-ter case, the Luddites thoughtfully discriminated regarding which machines were to be destroyed. As one account at the time in the *Leeds Mercury* reported:

> They broke only the frames of such as have reduced the price of the men's wages; those who have not lowered the price, have their frames untouched; in one house, last night, they broke four frames out of six; the other two which belonged to masters who had not lowered their wages, they did not meddle with."[15]

The Luddites were not antitechnology; they were concerned, as Thompson con-cludes, that "industrial growth should be regulated according to ethical priorities and the pursuit of profit be subordinated to human needs."[16] That surely strikes us as an admirable goal.

But what of the commonly held view, with its echoes in the present, that pro-test against progress is pointless, and that the efforts of the Luddites were futile? Was "the triumph of mechanization" inevitable, despite the fact that "all but a mi-nority of favoured workers fought against the new system"?[17] To these questions we have two responses, both of which contribute to the cultural studies approach to technological culture that we develop in Section III. First, it is incorrect to think that the Luddite movement was completely ineffective. While it certainly did not hold up the general advance of industrial capitalism, there were many small victories in which the voice of the workers mattered. For the most part, Luddism segued into legal, parliamentary forms, thus making it difficult to de-termine how influential the Luddite spirit was in the troubled political landscape after 1818. The Corn Laws, passed in 1815, which kept corn prices artificially high, thus literally starving the working classes, were eventually repealed after a protracted struggle. Other reform bills during the 1820s and 1830s helped to alleviate deplorable working conditions and to assuage working-class resentment to the extent that England did not have a revolution, as did other European coun-tries.[18] The efforts of the Luddites may have counted for something. Indeed, this is not a matter of the triumph of manufacturers versus the triumph of the workers. The role of workers in the evolving technological culture is never a "done deal," but an ongoing and changing relationship, within which the sites of and reasons for struggle shift dramatically. There have always been those who have argued for prioritizing ethics and human needs over profit; and their efforts, no doubt, have kept industrial capitalism from denigrating the life of workers more than it has. The Luddites exemplify the need to keep up the pressure.

Second, the Luddites provided a potent alternative to the concept of technology and culture in the received view, at a time when the received view was gaining acceptance. They knew from their daily experience that technology is never neutral, never merely a tool. They knew that technology is woven into the fabric of daily life and that it is to be judged in relation to the quality of everyday life. It is never automatically progress. They knew that what constitutes convenience for some might have undesirable consequences for others. Further, as their activities make clear—activities in which they risked their lives—they knew that the development and implementation of technologies was not inevitable, and that human choices and actions are shaped by conscious political interventions. It is unfortunate that so much of what else they might have to say to us has been lost in the vicissitudes of political power, that their voices were silenced, that they have not been taken more seriously. It is certainly within our power, however, to take seriously any lessons we have gleaned.

Contemporary Luddism

Along with growing concerns about the effects of unbridled technological "progress," and the revised understanding of the history of Luddism, *neo-Luddism* has become something of a contemporary rallying cry for a number of individuals and groups engaged in analyzing and/or resisting technology in some form or another. There is even a certain cachet attached to the claim of being a Luddite. Kirkpatrick Sale draws the parameters of neo-Luddism with a broad brush, "ranging from narrow single-issue concerns to broad philosophical analyses, from aversion to resistance to sabotage, with much diversity in between."[19] For example, Frank Webster and Kevin Robins conceptualize an analysis of information technology as "a Luddite analysis," which, for them, means that it "refuses to extract technology from social relations," and insists instead that technology "must be regarded as inherently social and therefore a result of values and choices."[20] In contrast to this more philosophic variant of neo-Luddism, "ecotage" of the kind sometimes practiced by groups like Earth First! and romanticized by Edward Abbey in *The Monkey Wrench Gang* also receives the imprimatur of the Luddite.[21]

It is important to remember that the Luddite movement was conjuncturally specific: It made sense within a particular historical moment, and that moment has passed. Today, those who claim to be neo-Luddites occupy a spectrum so broad as to guarantee little about their position beyond a willingness to challenge technological development in some form. Consequently, it does not provide a platform on which to build a response to technological culture that can take us very far. Accounts of the Second Luddite Congress and the Second Luddite Congress II illustrate the difficulty. Organized by Scott Savage, a plain Quaker in Ohio, the Luddite Congresses proposed to bring people together under the Luddite umbrella. As described by Nicols Fox, the Second Luddite Congress, held in 1996, was "contentious and emotional as well as inspiring.... There were loud voices of disagreement that left many participants upset. One source of the conflict was the idea that resistance must always take a nonviolent form."[22] While they had much in common in a broad sense, they had little in common on which to build an organized political movement. They were unable to use the legacy of the Luddites to bridge the extreme positions of the anarchist and the conservative Quaker. The best they could agree on was this—not very satisfying—draft statement:

> Technology is out of control and is unraveling society.
> We are in an ecological, social and spiritual crisis.
> The needs of people exceed the needs of machines.
> There are signs of hope—we are building bridges to people who are "in" the machine.
> We must stand with others, serve others, build community with all people and with all creation.[23]

The Second Luddite Congress II seemed unable to proceed with an even more diverse group, which ran the gamut from "Rasta locks" to "the narrowness of [Scott Savage's] religious perspective."[24] When Fox went in search of modern-day Luddites, she found:

> That what accommodations they make to civilization vary from individual to individual and from year to year. Sometimes the goal is to avoid certain technologies, sometimes it is independence, sometimes it is to live more lightly on the earth for environmental reasons. Other times it has nothing to do with the environment.[25]

The particular self-identified neo-Luddite propositions with which we have most sympathy in developing an alternative to the received view are those proposed by Chellis Glendenning in 1990. Summarized here by Sale, Glendenning resists the blind allegiance to progress, rejects the sense that technologies are neutral tools, and calls for critique that places technology fully within its cultural context. She calls for:

> 1. Opposition to technologies "that emanate from a worldview that sees rationality as the key to human potential, material acquisition as the key to human fulfillment, and technological development as the key to social progress."
> 2. Recognition that, since "all technologies are political, the technologies created by mass technological society, far from being "neutral tools that can be used for good or evil," inevitably are "those that serve the perpetuation" of that society and its goals of efficiency, production, marketing, and profits.
> 3. Establishment of a critique of technology by "fully examining its sociological context, economic ramifications, and political meanings...from the perspective not only of human use" but of its impact "on other living beings, natural systems, and the environment."[26]

We conclude, then, that we have much to learn from the Luddites about the possibilities of resisting progress blindly, about recognizing the political nature of technology, and about understanding and critiquing the integration of technology into everyday life. In Section III, we talk about this integration in terms of articulation and assemblage. However, it is important to recognize that Luddism, as a historical movement, must be understood within the historical conjuncture that made it a meaningful response. We can learn from the Luddites to keep asking important questions about contemporary technological culture; but the specific conjuncture within which we live requires responses crafted to address the present.

CHAPTER SIX

Appropriate Technology

APPROPRIATE TECHNOLOGY IS A DIRECT RESPONSE to the perceived failures of the widespread allegiance to, and application of, the received view of culture and technology on a global scale. Appropriate Technology rejects the idea and practice of large-scale, industrial megatechnology as indicative of progress; it rejects technological dependence in favor of autonomy; and it recognizes the integral nature of technology in the quality of everyday life. Unlike Luddism, discussed in the previous chapter, and the Unabomber, discussed in the next, the activities of appropriate technologists have the decided advantage of being legal, and the views and strategies of appropriate technologists are readily available for scrutiny.

Appropriate Technology (typically shortened to AT) refers to a particular kind of technology: that considered appropriate to achieving certain goals. It is also refers to a movement, akin in some ways to Luddism, that is concerned with making certain kinds of (appropriate) technological choices. It is, however, an even more diffuse movement than historical Luddism. Like any movement, AT is integrally related to the historical context within which it emerges: in this case at the nexus of the 1960s and 1970s counterculture, and the reactions against international development projects. It is a practice and a sensibility born of a particular era. While there are important lessons and strategies to be learned from it, its significant limitations necessitate the development of theory and practice beyond its confines.

Sources and Varieties of AT

AT comes in many forms with many different names: appropriate technology, alternative technology, intermediate technology, radical technology, small-scale technology, convivial technology, environmentally friendly technology, sustainable technology, energy-efficient technology, low-impact technology, soft technology, people's technology, liberatory technology, and so on.[1] The theme is apparent in the list of names: AT is about making technological choices that resist the development of technology at any cost. Instead, its guiding principal is to discern an acceptable or appropriate match between technologies and the structures of everyday life.

AT emerged in response to the proliferation of the ideas about development that we introduced in Chapter 1 on progress. In 1961, the United Nations passed a resolution declaring the "United Nations Development Decade: A Programme for International Economic Co-operation." Its objectives included:

> The achievement and acceleration of sound self-sustaining economic development in the less developed countries through industrialization, diversification and the development of a highly productive agricultural sector.[2]

As a consequence, the UN supported the introduction of a range of First World technologies into the Third World: technologies of power, such as dams; technologies of transportation, such as railways; technologies of communication, such as radio and television; and technologies of agriculture, such as tractors, fertilizers, and new hybrid seeds, in what was called "the Green Revolution."[3] As many people have pointed out, the Development Decade was, for the great majority, a failure, and the Green Revolution had only partial success.[4]

The problem was that technologies were introduced with insufficient attention to the role these technologies would play in the reorganization of everyday life. The disasters are mythic and include unfortunate events such as the 1980 explosion of the fertilizer plant located in a heavily populated area in Bhopal, India, and the marketing of canned milk to replace infant breastfeeding in poor areas in South America. But nowhere is the failure of development technology more dramatic than in the failures of the Green Revolution. Vandana Shiva, who as written a great deal about the consequences of the Green Revolution on women and peasants, summarized it this way:

> The Green Revolution has been a failure. It has led to reduced genetic diversity, increased vulnerability to pests, soil erosion, water shortages, reduced soil fertility, micronutrient deficiencies, soil contamination, reduced availability of nutritious food crops for the local population, the displacement of vast numbers of small farmers from their land, rural impoverishment and increased tensions and conflict. The beneficiaries have been the agrochemical industry, large petrochemical companies, manufacturers of agricultural machinery, dam builders and large landowners.... The "miracle" seeds of the Green Revolution have become mechanisms for breeding new pests and creating new diseases.[5]

While awareness of the failures of the Development Decade was widely shared from its onset, there seemed to be little alternative to it. Witold Rybczynski notes, "Even as advanced technology was criticized, it was apparent that it remained the only way to progress, and for most less developed countries, the only desired way."[6] There seemed, then, no real choice, even if that choice was a failure; the power of the received view seemed insurmountable. An alternative of some sort was needed.

In response to these failures of development technology, a group called the Intermediate Technology Development Group held a conference in 1968 in England. They called it the Conference on Further Development in the United Kingdom of Appropriate Technologies for, and Their Communication to, Developing Countries.[7] Spreading out from the work of members of this group and participants of the conference, and connecting with the larger sense that technology was out of control, the AT movement emerged. The founder and director of the Intermediate Technology Development Group, Ernst Friedrich Schumacher, widely known as simply E.F. Schumacher, is often considered the father of the AT movement. His book *Small Is Beautiful*, first published in 1973, is likewise considered its manifesto.[8]

While working as an economist and civil servant in Britain, Schumacher traveled to Burma and India, where his experiences made him question the focus on high technologies that he saw there. He acknowledged that there was a need for technology in the Third World, but noted that the imported high technologies benefited a small elite, and were of no use to the majority of the population. What they needed was to reorganize the workplaces in rural areas and small towns in response to their condition of being labor rich and capital poor. The overall task, as Schumacher saw it, was:

> *First*, that workplaces have to be created in the areas where the people are living now, and not primarily in metropolitan areas into which they tend to migrate.
> *Second*, that these workplaces must be, on average, cheap enough so that they can be created in large numbers without this calling for an unattainable level of capital formation and imports.
> *Third*, that the production methods employed must be relatively simple, so that the demands for high skills are minimised, not only in the production process itself but also in matters of organisation, raw material supply, financial, marketing, and so forth.
> *Fourth*, that production should be mainly from local materials and mainly for local use.[9]

According to Schumacher, the appropriate technology in this case would be intermediate; that is, "more productive than the indigenous technology...but it would also be immensely cheaper than the sophisticated, highly capital-intensive technology of modern industry."[10] AT should be more democratic than capital-intensive technology; it should benefit most of the people and not just the elites; and it should be culturally sensitive to the organization of everyday life. Therefore, it should avoid the disruptions that can be brought on by the introduction of new technologies. AT, according to Schumacher, was not a return to a "primitive" past; AT does not have to be simple or traditional. It can be, and often must be, created anew, and scaled to meet local needs and conditions in a sensitive manner.

It is noteworthy that another of the seminal works adopted by the AT movement was written by a thinker with vast international experience. Ivan Illich, the author of *Tools for Conviviality*,[11] was born in Vienna in 1926. After leaving there in 1941, he traveled widely until his death in 2002. He has been described as a "polymath and polemicist" whose work as a philosopher, Roman Catholic priest, and activist took him to Puerto Rico, Central and South America, and the United States. He founded the radical Intercultural Center for Documentation in Cuernavaca, Mexico, in 1961, which trained volunteers to work in Latin America. His work in the 1970s and 1980s focused on alternative versions of development, including schooling, economics, energy, transport, and technology.[12]

Illich, like Schumacher, objected to the imposition of high technology by experts, and was in favor of promoting technologies that he considered "convivial." He defined convivial thus:

> Tools foster conviviality to the extent to which they can be easily used, by anybody, as often or as seldom as desired, for the accomplishment of a purpose chosen by the user. The use of such tools by one person does not restrain another from using them equally. They do not require previous certification of the user. Their existence does not impose any obligation to use them. They allow the user to express his meaning in action.[13]

Convivial tools "give each person who uses them the greatest opportunity to enrich the environment with the fruits of his or her vision," a goal that, according to Illich, is denied by industrial tools.[14] Conviviality, for Illich, designates "the opposite of industrial productivity."[15]

Illich did not offer up designs for convivial tools, although he named some (motorized and non-motorized bicycles, power drills, mechanized pushcarts, and telephones); nor did he detail what a convivial society would look like. Rather, he recognized that, in part, some of the obstacles standing in the way of the coming of a convivial society are those of imagination. Simply put, it is difficult to imagine a transformation of this magnitude. What he did offer are tools for the imagination, criteria for discerning whether a tool is using a person or vice versa, and criteria for determining whether a system of technology fosters independence or dependence.

It is interesting to note that Illich's *Tools for Conviviality* and Schumacher's *Small Is Beautiful* were published in the same year. The awareness of the need for AT was clearly "in the air": not just because of developments in the underdeveloped, developing, or Third World nations, but also because of what was happening in North American culture. The 1960s and 1970s rise of AT coincides with the rising interest in social-responsibility movements and with the emergence of what has been called the counterculture. Barrett Hazeltine and Christopher Bull point out that many projects and groups formed that they consider social-responsibility groups, in which the goals of AT were embraced. Such groups included the National Appropriate Technology Center; the projects of President Jimmy Carter; the Office of Technology Assessment; and even the projects of the USAID, the foreign-aid division of the State Department.[16] But perhaps even more to the point is AT's connection with the developing counterculture.

During the 1960s and 1970s, a mélange of people (primarily youth), disenchanted with what they considered the "establishment," sought alternatives to the dominant culture. These people, known widely as the counterculture, focused on creating alternative political structures based on anticapitalist, anti-industrialist values such as personal growth, self-realization, self-expression, pleasure, and creativity. It's easy to see how this movement articulates to the AT movement, because AT, as it was understood, tended to be anti–big industry and pro-individual. Indeed, members of the counterculture carried around copies of Schumacher and Illich as if they were the maps they needed to make the world a better place.

There remains an active counterculture for whom the works of Schumacher and Illich serve as canonical texts. This group tends to identify more with environmental causes than it did during the 1960s, but it also increasingly identifies itself as Luddite. For example, in 1978 Theodore Roszak's book *Person/Planet: The Creative Disintegration of Industrial Society*, identified with the counterculture.[17] His 1994 book, *The Cult of Information*, identifies itself in the subtitle as *A Neo-Luddite Treatise*.[18] Similarly, when Nicols Fox goes in search of Luddites, as we discussed in the previous chapter, the people she designates as Luddite are indistinguishable from people most of us would understand to be members of the counterculture: They live lightly on the land, use alternative energy sources, don't work nine-to-five, don't watch television, resist succumbing to consumer culture, are anticapitalist, and so on.

Moving with and beyond AT

The contemporary articulation of AT, the counterculture, and neo-Luddism, is a complicated one. However, it does explain a lot about the strengths and weaknesses of AT. We can summarize the significance in this way:

1. AT's focus on tools complements the counterculture's emphasis on lifestyle rather than social change. Of consequence: Tools and lifestyle matter more than major cultural change.

2. The vague nature of the AT's criteria for appropriateness complements the abstract nature of both countercultural and neo-Luddite commitments. Of consequence: No consistent, theoretical approach for evaluating what is "appropriate" emerges that can match the complex cultural nuances of technological culture.

3. The counterculture's focus on the individual complements and underscores AT's focus on the human–tool interaction as the object of analysis. Of consequence: The consideration of what matters is too narrowly drawn.

We will unravel these insights for you.

1. AT tends to focus on tools, as if the structure of the tools themselves is what matters. There is an important insight for us to take from this:

Tools do matter; however, it is not only the tools that matter. Ironically, we learned that from AT! You can't put a technology developed in one context into another context, and expect it to perform in the same way. Context matters too. So when Illich writes, "I will focus on the structure of tools, not on the character structure of their users,"[19] he sets aside too much. This logic underscores—almost ironically—the counterculture's emphasis on lifestyle, as if how one lives is all that matters. According to this logic, one need not engage in politics on a larger scale; human scale is all that matters. The tools one uses are the measure of one's worth. We argue, instead, that what is needed, and what AT cannot quite give us, is a map for fully engaging the multiple layers of connections among the tools and the user, among the device and the larger social structure within which it occurs. Without that map, a technological politics will be severely limited. That is what we work toward in Section III.

2. Vague guidelines for what constitutes "appropriate" derail the potential of analysis to reveal complex nuances that are worthy of attention. In short, small is not always beautiful and convivial is not always desirable. Rybczynski tells a great story illustrating this difficulty. The story, drawn from the account of Finnish anthropologist Pertii J. Pelto, goes something like this: The Skolt Lapps of northeast Finland adopted snowmobiles to make the difficult task of herding their reindeer easier. The consequences of this have turned out to be considerable. The community changed markedly as a number of realities changed: Mechanized herding gave younger, less-skilled men an advantage they never had before. Herding could be done in much less time, freeing up time for other activities. Easier travel facilitated more socializing. The cost of maintaining snowmobiles increased financial pressures. A new social stratification emerged based on who owned or who did not own snowmobiles. And most interesting, all this has changed the relationship between the Lapp and the reindeer. Because snowmobile herding is stressful to the reindeer, the health and size of the herds may be compromised. But even more significant, where the relationship used to be proximate—based on the ability of skilled Lapps to tame their reindeer—the spatial and psychic distance has increased dramatically. The relationship of man to reindeer has been transformed.[20] So much has changed, and perhaps not all for the better. Yet, by most criteria of AT, the snowmobile is an appropriate technology: "it is small, easy to operate and maintain, encourages decentralization, and is not very expensive." Furthermore "it was not imposed but freely chosen."[21]

    What we can see in this example is the fact that the abstract nature of the criteria for appropriateness is not enough to really understand the complexities of technological culture. The search for, and satisfaction with, such criteria make the hard work of understanding seem easier than it really is. The appeal of this kind of logic to those who claim to be neo-Luddite or countercultural is obvious. They are criteria that are hard to argue with. They lend support to the neo-Luddite sensibility and the countercultural emphasis on the individual relationship with one's tools. But they do not insist on thinking through the nature of complexity, as illustrated by the case of the Lapps. Again, attention to that kind of complexity is essential to the approach we propose in Section III.

3. When the focus in on the individual, as is the tendency in the counterculture, technologies will be measured in terms of what they do for the individual, and not much beyond. This tendency lends support to AT's human-tool orientation as the object of analysis, and promotes a kind of blindness to larger questions about what might matter. We illustrate this complementary work with an example from Illich, who claims that the telephone is a convivial tool. Why? "Anybody can dial the person of his choice if he can afford a coin.... The telephone lets anybody say what he wants to the person of his choice; he can conduct business, express love, or pick a quarrel. It is impossible for bureaucrats to define what people say to each other on the phone."[22] The analysis stops here, having satisfied the criteria for what makes a tool convivial and giving support to the notion that what matters is cheap, unfettered communication among individuals of one's own choosing. If one has the perspicacity to look beyond the satisfaction of the individual, this characterization of telephone technology is woefully incomplete. What of the structure of ownership of the telephone industry? Who benefits financially? Who does and who does not have access to a telephone? Who does and who does not have those few coins to make the call? What role does the telephone play in the spatial organization of family and friends? What about telephone lines and cables, competition, investment, the use of the telephone for surveillance, and on and on? There is simply so much more to consider beyond the individual act of picking up the phone and being free to talk to anyone. Again, in Section III we point the way to making sure that all those larger questions are part of how we understand technological culture.

With those limitations in mind, we conclude by acknowledging the enormous debt we owe to the AT movement. AT does challenge the blind allegiance to progress. It does insist on cultural sensitivity. It does strive for something quite admirable, which we wish to take along with us. That is, as Hazeltine and Bull put it, the concept "that the technology must match both the user and the need in complexity and scale."[23] We just want to think more broadly about the kind of complexity we consider, assess the concept of needs beyond the human–tool interaction, and expand scope of understanding beyond the individual or small group.

CHAPTER SEVEN

The Unabomber

Bᴇᴛᴡᴇᴇɴ 1978 AND 1995 A MAN THE FBI referred to as "the Unabomber" mailed a series of bombs to universities and corporations across the United States, resulting in the deaths of three men, and the injury, some serious, of twenty-three others. He was referred to as the Unabomber because his victims seemed to be related either to academia (university) or the airlines industry—thus, Un-A-Bomber. The victims were, for the most part, not major public figures. As Tim Luke has described them, they were part of a new class of "comparatively obscure administrators, agents, or academicians who were actively working in the applied sciences, computer sciences, or mathematical sciences for small firms or universities."[1]

In 1995 the bomber, referring to himself as "FC," which stood for Freedom Club, an organization to which he said he belonged, began writing public letters to individuals and newspapers. He expressed his frustration with the crushing alienation of industrial society. Later that year he offered to cease his bombing campaign if major newspapers would publish his 35,000-word essay expressing his views, and two 2,000-word essays, one each in subsequent years. On advice of the FBI, the *Washington Post* and the *New York Times* reluctantly published the essay titled "Industrial Society and Its Future," which was quickly dubbed the Unabomber's "Manifesto."

David Kaczynski read the essay and recognized in it key ideas and phrases similar to those that his brother had used in letters home. He related this information to the FBI, which subsequently arrested Theodore ("Ted") Kaczynski for the Unabomber's crimes.

We discuss the case of the Unabomber here because he represents the most extreme contemporary critique of technological culture.[2] Also, the Unabomber's Manifesto, "is the most widely circulated writing in the field of science, technology, and society" because of the notoriety and circumstances of its publication.[3] Kaczynski has also become something of a myth: the insane hermit, the nut in the woods, a mythic archetype who resonates strongly with militia and survivalist groups who likewise reject society and take up armed resistance. He seems to fit within a frightening trend in society that has increased in the past two decades, a trend toward isolation and violence.[4] What is especially disturbing about the Unabomber case is that many of the critiques of industrial society espoused in "Industrial Society and Its Future" are ones that we have written about ourselves, assigned as class reading, and consider to be classic statements in the field of technology studies. If we condemn the essay and its ideas in their entirety—as the work of a madman, as many are wont to do—then we will also have to condemn Jacques Ellul, Lewis Mumford, Herbert Marcuse, Ivan Illich, and many others. As Scott Corey has argued, there has been a profound silence on the part of academics in responding to the Unabomber case—despite the fact that the essay is assigned as reading in classes across the country—perhaps because Kaczynski hits too close to home.[5] If we dismiss him as an irrational nutcase (as many have dismissed the Luddites), we do not need to recognize or engage with what rings true, or at least what merits consideration in his ideas.

In this chapter, we first discuss the insistence that Kaczynski is insane. Second, we view the despairing picture of a totalizing industrial society that drove him to commit the acts that he did.

Kaczynski Must Be Insane

The arrest of Theodore Kaczynski was an event for the media. The Unabomber's seventeen-year reign of terror ended when the FBI raided the one-room cabin in Montana where he lived a hermit's life. In all the press photos, Kaczynski looks wild-eyed, with long, unkempt beard and hair. He had been living for decades in self-imposed low-tech conditions, growing or hunting most of his food, and having no running water or indoor sanitation facilities. As his background was uncovered, he was shown to have been a brilliant mathematician who entered Harvard at age sixteen and taught at the University of California at Berkeley before heading for his cabin in the woods. It was easy for the press to brand him as an extreme loner, a boy genius who had gone insane, existing far outside of society. Portraying him in these terms made it easy to dismiss Kaczynski as an aberration. This portrayal as an extreme loner served to disconnect him from anarchist movements, environmental movements, or even a long tradition of the critique of technological culture. In a way, the public needed him to be a loner so they would not have to worry about him anymore after his arrest and trial.

Although he was portrayed as insane, his insanity was never proven. As Alston Chase explains, most diagnoses of Kaczynski's insanity came from two forms of analysis.[6] First, diagnoses were based on superficial analyses of his lifestyle. Thus, to live alone, to live without much twentieth-century technology, to be celibate, to be misanthropic, and to be a loner is to be insane. Second, diagnoses were based on examinations of his writings, which are inadequate bases for a genuine diagnosis of insanity. Some claimed that he was insane because he did not admit that he was insane or would not cooperate with experts who wished to declare him insane; a Catch-22 if ever there were one! To admit to being insane is to be insane; but to deny being insane is also to be insane! Even Kaczynski's own lawyers, without his knowledge, based their case on an insanity defense. When Kaczynski found out, he tried to fire his lawyers, and failing that, asked to represent himself in court. The only court psychologist to examine Kaczynski in response to his own request, Sally Johnson, concluded that Kaczynski was competent to stand trial and to represent himself in court. She gave a provisional diagnosis of "paranoid schizophrenia," but apparently did not think that this hindered his competency. The judge still refused to let Kaczynski represent himself, which, many suspected, would have led to a very public and political trial. Therefore, Kaczynski accepted a bargain to plead guilty and spend life in prison rather than face a trial in which he would have been presented as insane.[7]

Actually, Kaczynski was not quite the hermit and loner that most have portrayed. He traveled, read widely, and engaged in intense correspondence with many people throughout his time in Montana. This correspondence continued from prison.[8]

The Manifesto itself has been presented alternatively as the ramblings of a madman or a work of genius. Kirkpatrick Sale placed him within a long line of neo-Luddites about whom Sale was just finishing a book, which we referred to in Chapter 5.[9] But there is a reluctance to look at the Manifesto too closely, or to critique it on its own terms. Perhaps this is because if one were to take his work seriously, even if aiming to discredit each of his arguments, one would have to acknowledge places where FC has a point about the technological nature of society and its restrictions on free will.[10] To agree with any part of the Manifesto might be seen as agreeing with FC's conclusions and methods (justifying his terrorism and murder).

But this need not be the case. One can engage and even agree with points that FC makes without advocating murder or violence. In fact, two quite detailed and insightful critiques of the essay—by Tim Luke and Scott Corey—have appeared in recent years.[11]

The Unabomber Manifesto

"Industrial Society and Its Future" is a fairly well-organized essay. Aside from its notable digressions against "The Left," it warrants a closer look. Its argument is set out in the opening paragraphs:

> The Industrial Revolution and its consequences have been a disaster for the human race. They have greatly increased the life-expectancy of those of us who live in "advanced" countries, but they have destabilized society, have made life unfulfilling, have subjected human beings to indignities, have led to widespread psychological suffering (in the Third World to physical suffering as well) and have inflicted severe damage on the natural world. The continued development of technology will worsen the situation. It will certainly subject human beings to greater indignities and inflict greater damage on the natural world, it will probably lead to greater social disruption and psychological suffering, and it may lead to increased physical suffering even in "advanced" countries.

The industrial-technological system may survive or it may break down. If it survives, it MAY eventually achieve a low level of physical and psychological suffering, but only after passing through a long and very painful period of adjustment and only at the cost of permanently reducing human beings and many other living organisms to engineered products and mere cogs in the social machine. Furthermore, if the system survives, the consequences will be inevitable: There is no way of reforming or modifying the system so as to prevent it from depriving people of dignity and autonomy.[12]

Roughly, the argument of FC is that technological society works for its own ends and not for the real needs of the individual. The individual is shaped to meet society's needs, and not vice versa. All aspects of modern society work to dehumanize and disempower the individual. The industrial system is due for a collapse, and the more humans are dependent on that system (and most are radically dependent on it), the harder that crash will be. It is FC's goal to bring about the earlier rather than the later collapse of industrial society. He aims to bring humankind back into balance with nature and with personal autonomy, where individuals or small groups can exist without being subordinated to corporations, bureaucracies, or any other system. Modern technologies are so thoroughly permeated with power and domination that they cannot be rearticulated for other democratic or libertarian uses. They must all be destroyed, and all the technical manuals burned.[13]

According to FC, industrial society and its future is marked by absolutes: "technicism" has penetrated all aspects of society and nature absolutely; technology and "wild nature" are in absolute opposition to one another; small-scale society and small-scale technology are absolutely good; large-scale society and large-scale technology are absolutely evil. There is no compromise and no possibility of compromise. Those who compromise are part of the problem, and since there is no compromise, FC sees no other solution than the path he has taken. As Luke explains:

> No vocabulary is fully adequate for reiterating what the Unabomber attacks in his manifesto or for explaining how someone could commit this sort of violent action. On one level, it is about power and knowledge turning an individual against technoscientific structures because of the frictions felt by all individuals living within industrial, bureaucratic society. On another level, it is a plea to recollectivize people and things on a smaller scale, at a slower pace, and in simpler ways. And on a third level, it is a shallow justification for mayhem and murder.[14]

As Corey describes, Kaczynski was profoundly influenced by the work of Jacques Ellul, especially Ellul's groundbreaking book, *The Technological Society*, in which Ellul argues that modern society is characterized by all-encompassing *technique*, which permeates all aspects of modern life.[15] So deep is the reach of technique that the only escape from such a society is radical catastrophe or the intervention of a loving God (Ellul was a priest). However, Ellul never advocated violent rebellion and even thought that political action was useless. Kaczynski wrote letters to Ellul in the early 1970s, though it is unknown if Ellul ever responded or even read them. What we do know is that by 1976, in his book *The*

Ethics of Freedom, Ellul had "denounced virtually every FC position."[16] Kaczynski refused to acknowledge these dimensions of Ellul's work.

FC's rhetoric is shaped by the intellectual and social climate of the 1950s and 1960s, especially the idea that humankind is becoming dominated by an all-encompassing system. Such ideas were prevalent in William H. Whyte's *The Organization Man* and Herbert Marcuse's *One-Dimensional Man*, both of which argued, in different ways, that humanity was being reduced by the focus on the practice of obedience to authority, the value of efficiency through technology, and the overarching goal of corporate profit.[17] Taken out of the immediate context of their publication, many aspects of FC's argument would have found their place in courses on technology and culture for the past 30 years. For example, consider a paragraph from Lewis Mumford's *The Myth of the Machine: The Pentagon of Power*:

> The business of creating a limited, docile, scientifically conditioned human animal, completely adjusted to a purely technological environment, has kept pace with the rapid transformation of that environment itself: partly this has been effected, as already noted, by re-enforcing conformity with tangible rewards, partly by denying any real opportunities for choices outside the range of the megatechnic system. American children, who, on statistical evidence, spend from three to six hours a day absorbing the contents of television, whose nursery songs are advertisements, and whose sense of reality is blunted by a world dominated by daily intercourse with Superman, Batman, and their monstrous relatives, will be able only by heroic effort to disengage themselves from this system sufficiently to recover some measure of autonomy. The megamachine has them under its remote control, conditioned to its stereotypes, far more effectively than the most authoritative parent. No wonder the first generation brought up under this tutelage faces an "identity crisis."[18]

The themes of the transformation of humans to the needs of the machine, the transformation of the environment, and the destruction of human dignity and autonomy are all themes in FC's manifesto. The difference is that neither Mumford, nor Ellul, nor Whyte, nor Marcuse killed anyone or advocated killing people as a viable solution to society's technological troubles. FC would undoubtedly argue that these figures simply lacked the courage of their convictions, and that they were all a part of the too comfortable, academically ensconced "Left."

FC presents two major arguments that might explain his violence. One is that FC felt that violence was the only way to gain attention:

> If we had never done anything violent and had submitted the present writings to a publisher, they probably would not have been accepted. If they had been accepted and published, they probably would not have attracted many readers, because it's more fun to watch the entertainment put out by the media than to read a sober essay. Even if these writings had had many readers, most of these readers would soon have forgotten what they had read as their minds were flooded by the mass of material to which the media expose them. In order to get our message before the public with some chance of making a lasting impression, we've had to kill people.[19]

FC's conclusion is obviously untenable, based on a number of unsupported assumptions. The most fundamental of these assumptions is that people would actually read the copy of the manifesto published in the newspapers, whereas it was most likely, as Luke surmised, "tossed away with the rest of the September 19, 1995, newspaper."[20] But if gaining attention was FC's sole objective, then he most likely would have targeted higher-profile people for his bombs.

The Unabomber's turn to violence is more likely the result of his despair at what he sees as the totalizing nature of *technicism*. We might define technicism as the imperialism of the received view of culture and technology as applied to the whole of human experience, or the methodological application of a technical logic to what is not technical.[21] The totalization of technicism, its intervention into every aspect of life, society, and nature, is a vision he draws from Ellul (but his conclusions are rejected by Ellul, as we said above). Both FC and Ellul would agree with Mumford's identification of a fatalism characteristic of modern society:[22] the unquestioning acceptance of technology as the only true path to improve humanity's lot, a "technological compulsiveness: a condition under which society meekly submits to every new technological demand."[23] What is key for Mumford is that this sycophantic attitude toward technology (and technocrats) is the product of a particular historic period and not innate to human beings. He points out several examples where scientists argue that it is simply human nature to pursue any technological or scientific possibility, no matter how destructive. This argument allows scientists, engineers, and technologists to completely ignore ethical and moral action. The problem with the fatalists, for Mumford, is that they simply cannot see any way of changing or reversing the seemingly inevitable conclusion of the technicist logic that they have accepted as gospel truth. As a historic phenomenon, this mind-set can be opposed (and must be opposed), and he points to a contemporary "affirmation of the primal energies of the organism" that he sees in the counterculture of the 1960s.[24] But rather than valorizing the counterculture, he warns that such forces can be just as destructive to humanity if left unchecked.

Chase argues that the roots of the Unabomber's actions lie in the atmosphere of despair and desperation of the 1950s and 1960s.[25] Though these forces did have, and continue to have, an influence on a whole generation, they are simply not an excuse for murder. The totalizing vision of technicism has found a ready audience again in the new millennium as information technologies record and control our lives in ways that Mumford foresaw but in areas and scope that he could not have foreseen.[26] It is tempting to see overwhelming forces (for example, individual consumers being no match for giant multinationals with their corps of lawyers) as absolute because then it seems to give one's struggle a moral force.

There are other violent responses to industrial society that approach the totalizing rhetoric of FC. Some base their actions on an appeal to religious grounds (such as the terrorism of 9/11 or the arming of fundamentalist groups such as the Branch Davidians), some to ecological grounds (such as some of the factions of Earth First!), and others to political grounds (anarchist groups and libertarian survivalists). The cry from many such groups, articulated especially well by the

Unabomber, is that there is no alternative and that industrial society is all encompassing: that modern technology is so thoroughly permeated by relations of domination and dehumanization that the only solution is society's destruction.

Lessons to Learn

To us, the death of billions of people in the collapse of global industrial society is completely untenable and immoral, as a goal or a solution. Likewise, we decry risking the death or injury of anyone to make some political point. Like Lewis Mumford, we do not hold with the universalizing view of the dominance of industrial society, and we need to work hard against the despair that such visions cause. But we also do not pretend to ignore the ways in which technicism has permeated everyday life.

We are not naive in our faith in resistance, nor do we tout resistance as yet another inevitable feature of technological society. One of the purposes of this book is to provide readers some of the tools for recognizing the important cultural dimension of technology, the important technological dimension of culture, and to examine the effects and possibilities for both human and technological action in our everyday lives. To this end, we must be able to recognize what is legitimate in the Unabomber's complaint, but incorporate it into a world view that better understands the possibilities for human action that reside in the complexities of technological culture. That is why, in Section III we propose a cultural studies approach to technology, which draws on concepts of agency, articulation, identity, space, and politics. It is through these concepts that we can envision a more constructive path than the deadly alternative of the Unabomber.

One possible outcome of the Unabomber story is this: Perhaps in those moments of recognition, when, in FC's writings, we glimpse ourselves as academics, Luddites, political advocates, or environmentalists—in the moment before we look away, shut him up, and drive him and his arguments from our consideration—perhaps those moments might be profoundly disturbing enough for us to fundamentally reassess what it is we really want, and how we want to get there.

There is evidence that groups of anarchists and Greens are moving away from the romance of violence and toward the building of broad coalitions of peaceful protest. While the media presented images of the World Trade Organization protests of 1999 that focused on violence, that violence was not inherent in the tactics and motivations of the thousands of people who gathered in Seattle to have their voices heard by the world. Rather, the violence was perhaps a symptom of how dangerous it is to even question what it is we really want and how we want to get there.

Electric Oven
Photograph by Theodor Horydczak, ca 1920–1950
Library of Congress
Horydczak Collection

Section III

A Cultural Studies Approach to Technological Culture

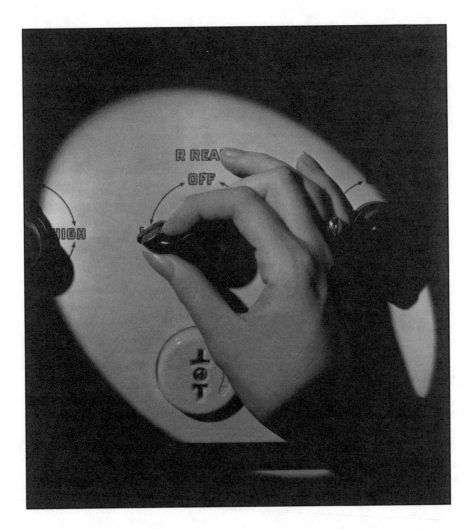

CHAPTER EIGHT

Defining Technology

W HEN YOU PAY ATTENTION to what people talk about—in casual conversations, in class, on radio and television, in books, and in films—you note that
they are often talking about, writing about, thinking about, reacting to, or responding to technology. Many of these conversations involve life-giving, life-changing,
and life-threatening matters; controversial topics include stem-cell research, genetic engineering, media surveillance, execution by the electric chair, the impact
of television on violent behavior, Jack Kevorkian's suicide machine, global warm-
ing, and weapons of mass destruction. Technology clearly matters, and it matters
enormously. In less dramatic ways, the topic of technology also pervades talk about
what matters in everyday life: in discussions of the breakup of Microsoft, new
film animation techniques, the development of hybrid cars, Mexican trucks on
US highways, or even in discussions about purchasing a digital camera, computer,
printer, cell phone, DVD player, PDA, and so on. Sometimes the matters seem
relatively trivial: such as expressions of frustration over malfunctioning answer-
ing machines, ATMs that are out of service, and gas-guzzling SUVs. Sometimes
we know that these matters are deadly serious: such as recent debates over which
countries can legitimately develop "weapons of mass destruction."

What is amazing about these conversations involving technology is how little
agreement there is about what is at stake, that is, about what really matters. This is
often dramatically the case when the topic is controversial. Consider, for example,
the controversy involving US physician Jack Kevorkian's machines that hasten
death. These simple machines are of two types. One is a set of intravenous bottles
mounted on a metal frame with a mechanism that allows the patient to turn on
and trigger the flow of a series of drugs that will bring on death painlessly. The
other is a tank of deadly gas and a mask with a mechanism that allows the patient
to turn on and trigger the flow of gas that will similarly bring on death painlessly.
Kevorkian and his machines have been the cause of considerable public and legal
controversy. Is Kevorkian a passionate physician or a cold-hearted murderer?

Some people argue that the machine honors a person's right to take control of his or her life and death. They believe that when people have experienced prolonged suffering, they ought to have the right to cease that suffering. From this perspective, Kevorkian is a virtual saint bucking an uncompassionate legal establishment, and his machines are "assisted-suicide machines," a compassionate way to help people gain control that would otherwise be denied them.

Other people argue that no human has the right to determine the moment of a human death, even one's own. Some fear the possibility that, once allowed to kill legally, the machine will surely be used to justify killing those who are deemed undesirable—in the manner that fascist Germany used liberal euthanasia laws to justify killing Jews, Gypsies, homosexuals, and the handicapped. To legalize Kevorkian's machines would be to invite fallible humans—and eventually the state—to kill at will. From this perspective, Kevorkian is an agent of encroaching totalitarianism, and his machines are "killing machines," an evil that will usher in legalized, political murder.

You may have heard or even participated in conversations where these (or variations of these) arguments about Kevorkian's assisted-suicide or killing machines have been made. What all these conversations have in common is attentiveness to the fact that this technology clearly matters, and it matters enormously (either in a positive or negative way). What these conversations often do not have in common is agreement about what is at stake, or what matters: Does the individual have a right to choose the time of his or her death? Do states have a right to murder those deemed undesirable? These discussions often end frustratingly, at an impasse, without a way to reconcile what are seen as mutually exclusive stakes. There is seldom a shared framework for deciding, among the many decisions that might need to be made, if the machines should be legal or illegal.

This problem is enacted daily, at every level of conversation concerning technology, even at the most mundane level. For example, in discussing the desirability or undesirability of SUVs, what exactly matters: that there are too many polluting automobiles on the road? That people have the right to drive whatever they want? The excessive stress put on the environment due to overpopulation? The restrictions on domestic drilling that limit the availability of gas? The rollover rate of SUVs? The tendency of SUV drivers to fare better than the drivers of smaller vehicles in crashes between them?

In conversations about these topics, the reason we fail to reach more constructive outcomes can be understood partly in terms of a very significant lack: the lack of a sophisticated and shared understanding of how to approach questions of technology. Even if all participants agree (explicitly or implicitly) to consider that the matter in question is technological, it is striking how little agreement there is about precisely what that means. What exactly are people talking about when they support or criticize the existence of Kevorkian's machines or SUVs? What, after all, *is* technology, and how is it connected to our assessments of all the other aspects of daily life that matter? Without that key, that sense of common theoretical ground, we remain destined to discuss, argue, and live at cross-purposes in a communicative space where we cannot begin to sort out the basic terms of disagreement. Without that key, our mechanisms for achieving resolutions to technological matters of enormous importance remain hopelessly flawed.

So, Then, What Is Technology?

Part of the difficulty with reaching common ground in discussions concerning technology is that the term is used in so many different ways. One could turn to the dictionary, but dictionary definitions do not adequately capture the meanings of technology that people operate with in everyday life. If you take a group of people and ask each person to write down a definition of technology, you will get as many definitions as there are people in the group! This is often the case even when they are allowed time to consult sources (such as dictionaries) or experts. There do tend to be, however, some thematic similarities in the definitions people turn up. Here are some typical definitions. Drawing on *Webster's*, technology is:

> 1 a : the practical application of knowledge especially in a particular area: ENGINEERING 2 <medical technology> b : a capability given by the practical application of knowledge <a car's fuel-saving technology>
> 2 : a manner of accomplishing a task especially using technical processes, methods, or knowledge <new technologies for information storage>
> 3 : the specialized aspects of a particular field of endeavor <educational technology>[1]

Rhetoricians typically define technology by pointing to the Greek root, *tekhne*, which means art or craft. The suffix *ology* means "the study of." When you put these two together, technology means the study of an art or craft. Cultural theorist Raymond Williams, in *Keywords: A Vocabulary of Culture and Society*, writes that technology is used to "describe a systematic study of the arts...or the terminology of a particular art" and has had this meaning since the seventeenth century.[2]

Interestingly, few people still make everyday use of the term technology in any of the above ways (if they ever did!). What is curious about these definitions is that they treat technology as application, capability, manner of doing, and specialized aspect, but not as a thing. When technology is referred to in popular discourse, however, it is almost always as things (tractors, pacemakers, computers, and so on). Even more interesting then is the fact that the examples in the dictionary definitions suggest things: medical technologies (e.g., respirator), fuel-saving technologies (e.g., catalytic converter), information storage technologies (e.g., computer), and educational technologies (e.g., computer set up for language instruction). In our estimation, the most common meaning of technology in popular usage conceives technologies as things that are useful; that is, as things that have, as the dictionary puts it, some "practical application." So technology is, at least in terms of its most popular usage, a constructed and useful thing.

What does it mean to treat technology as a "thing"? Or, as we prefer to think of it, in terms of its "thingness"? It means to understand and treat technology in terms of objects that have discrete boundaries precisely delimiting the objects and differentiating them from others. So, for example, a digital camera is a different technology than a film camera. Although they are related in some ways, it is possible to specify what makes each unique. Likewise, it is possible to differentiate technologies from other kinds of things. In this way of thinking, technologies (the camera for example) and culture are separate things, each occupying its own separate space. Although they may have a relationship, they are each separately bounded and definable. A technology may exist in culture, but like an egg in a nest, it is an isolatable, discrete object. A technology may touch, but not interpenetrate the other object: culture. Where one begins and the other ends is always decidable, a mere matter of calculation, measurement, and discernment.

Most often, technological objects are understood to be constructed, solid, and nonliving (although biotechnology is changing this somewhat). They are understood to be stable masses, that is, particular arrangements of matter that can be described in terms of their mass (large, small, heavy, light, soft, hard, dense, and so on). Technologies are artifacts, instruments, tools, machines, structures, and constructions; they are detached and different from living bodies and from other things. In this sense, they are discrete, isolatable objects, correlates of natural objects, but not natural. Examples of such things include cameras, paperclips, scissors, generators, automobiles, bridges, buildings, computers, televisions, overhead projectors, microscopes, CD players, CDs, and assisted-suicide/killing machines.

Thingness, however, also points to the fact that people often treat arrangements without solid mass as technologies—as things. An excellent example of this idea of technology is the Internet. While commonly thought of as a technology, the Internet does not occupy space in the same way that a computer monitor does. It is still commonly treated, however, as though it had a discrete, isolatable nature. Although the work of discernment is more difficult, it is possible to map its boundaries, to delimit what is the Internet and what is not. It is a network that consists of certain components of hardware, software, and certain more ethereal components such as electrical connections, microwaves, and satellite links. It is not the computer monitor, the user, the software or hardware designers, or the companies that post Web pages. It is, rather, the network of connections among these (and other) sites. Note: not the sites themselves, but the network of connections among them. Thus, even though the Internet has no "weight" (or other such definitive measure of mass), it is a constructed, nonliving, arrangement that is contained by boundaries that define what it is and what it is not. It has an inside and an outside. While it is a complex network, it does not interpenetrate the other "things" that make up the rest of culture.

The cultural tendency to conceive of technology in terms of thingness has interesting and serious consequences. Significantly, as we have argued, it directs vision toward the "stuff" of technology, the solid, measurable things that are produced. In so doing, it deflects vision away from the interdependent relations among the living and nonliving within which these things are given form. To focus on bounded artifacts—on thingness—is to deflect understanding from the ongoing energies, activities, relations, interpenetrations, and investments within which these things appear, take flight, and have effects. Further, the formulation of technology as things that are useful deflects vision toward the tool-like use of these things, and away from the work or role of these things beyond matters of their usefulness.

In the remaining chapters of this book, we develop a way of understanding technology that foregrounds the interconnectedness within which things appear, are developed, and have effects. While the approach we develop relies on the theoretical concepts of articulation and assemblage, it owes a great debt to many scholars who have proposed alternative approaches to conceiving the interconnectedness of technological culture. For example, in his book *Technology as Symptom and Dream*, Robert D. Romanyshyn defines technology as "an enactment of the human imagination in the world."[3] Andrew Feenberg, in *Critical Theory of Technology*, defines it as a "process of development suspended between different possibilities."[4] Langdon Winner, in *The Whale and the Reactor*, defines technologies as "forms of life."[5] Elizabeth Grosz has recently put it particularly elegantly. She writes in her article titled "The Thing": "Technology is that which ensures and continually refines the ongoing negotiations between bodies and things, the deepening investment of the one, the body, in the other, the thing."[6]

While these formulations may not yet make sense, they do point to flows, connections, and interpenetrations among the living, the nonliving, producers, users, processes, possibilities, and energies—and not just to things. If we can learn to think with definitions such as these, we may be able to find productive common ground from which to speak about technological culture.

Why Struggle with Meaning?

There are several forms of resistance that you might be feeling to this call to learn a new—and decidedly more complicated—sense of technology. First, you might ask, with all the definitions of technology available, why propose another? Wouldn't it make sense to simply advance the one that is "correct" or "best" and move on? Second, you may have a rather well-worked-out definition of technology with which you are satisfied. Perhaps you feel it has served you well up to now and see no need to abandon the comfort it offers. Third, you may challenge the idea that anyone has the "right" to simply develop (or "make up") a new definition as they see fit. You may believe that language and meaning are more fixed and absolute than to permit such tinkering. As we argue below, grappling with the problems of what technology means, and the power that different definitions have, actually provides crucial insight into the character of technological culture.

First, in response to the hope that we could simply choose the most comprehensive and useful definition of technology and move on, we maintain, as we have argued above, that there is no definition of technology that (as yet) seems to work consistently in everyday life. Dictionary definitions don't match up very well to actual use, and popular usage is inconsistent. Working to develop a widely shared, sophisticated understanding of technological culture might help us solve significant problems involving technology. But, in the interest of achieving that understanding, we can't simply jettison all the meanings and definitions of technology that have come before and that are a part of our culture. However inadequate or problematic they may be, they influence current understandings and actions—usually in inconsistent and contradictory ways. In a very real sense, all those definitions contribute to the shape of technological culture.

Second, in response to those who are comfortable with a particular definition of technology, we encourage you to put your definition to the test, in light of what you've read thus far in this book. Has it always served you well, or have you had to change your concept of what technology is from time to time in order to grapple with the issues that have been raised here? We suspect that the latter is the case. Why? Technology is—and will likely continue to be—polysemic. *Polysemy* is a term that points to the fact that words can have many different meanings. The more potential meanings that can be attributed to a word, the more polysemic that word is. Some words, at particular historical moments, are highly polysemic. Terms such as *love, democracy, freedom,* and *technology* are currently highly polysemic terms in North American culture. An understanding of the work performed by the term "technology" should be broad enough to accommodate the fact that technology is likely to remain polysemic, for it is a site of significant cultural struggle and change.

Third, in response to skepticism you might have about our "right" to develop a definition of technology, we next explore a little bit about the nature of language and meaning, to clarify that change, not stasis, is more the rule than the exception.

Struggles over Meaning

Most people are familiar with the distinction between denotation and connotation. Denotative meaning implies that a word has a precise, unambiguous, or correct meaning. A word, in this case, signifies, or denotes, an explicit and culturally shared meaning. If, for example, you want the denotative meaning of the word technology, the best source is the dictionary, which delivers the "real" meaning. It is interesting how often students writing papers on controversial topics will go—naively—to the dictionary for the "real meaning" and hence the "final word" on some topic, as though the dictionary was the final authority on what something "really is."

The dictionary, as we discussed above, is not the best place to look for the meanings of technology (or many other terms) used in everyday life. For that you need to understand connotative meanings: meanings that are implied by a word, meanings that are, in a sense, lived. Connotative meaning refers to the fact that words imply or evoke associations, memories, commitments, values, beliefs, and affects. These meanings are harder to track down than are denotative meanings, because they tend to be less consensual, less culturally explicit, and less likely to be "codified" in dictionaries. For many people, technology connotes progress; they encounter the word with enthusiasm, participating in a belief that new technologies make our lives better. For others, technology connotes economic hardship; they encounter the word with dread, believing that technology refers to the expensive things in life they would like to have but cannot afford, or to the objects responsible for the loss of a job. Connotative meanings such as these vary dramatically, because they point to different—and often highly complex—ways of living in and experiencing the world.

Although connotative meanings are more difficult to assess than denotative meanings, they often play the more powerful role in everyday life. This is clearly the case with technology, where, as we stated above, almost nobody actually uses or lives with the denotative dictionary definitions. As a result, it is a rather difficult to track what the powerful connotative definitions are, and what cultural effects those definitions have.

This task is made more difficult by the fact that meanings change—even denotative definitions—and that there is traffic between denotative and connotative definitions. In actuality, the distinctions between denotation and connotation are not absolute. Language, after all, does change, and dictionaries—to some degree—reflect those changes. New meanings develop in a culture and sometimes make it into the dictionary. For example, you'll find "Internet" only in a fairly new dictionary. Further, old meanings sometimes disappear. *The Oxford English Dictionary* is a resource that specializes in tracing the changing meanings of words. The changing meanings are significant because they demonstrate that no denotative meaning is absolutely "true." Rather, meanings are true—perhaps temporarily—simply because there is wide cultural agreement on a meaning and lexicographers have chosen to put these meanings in their dictionaries.

In a sense, then, all meaning is connotative. All meanings are implied, subject to change, and liable to be legitimated (or not) in a complex process of cultural change. At different historical moments, different meanings will seem more or less contested, because, we remind you, there is often very much at stake in how you define something. It truly does matter, for example, whether you define Kevorkian's machines as "killing machines" or as "assisted-suicide machines." If you wanted to use one of these machines to terminate your life, it would matter. A killing machine might not be legal or easy to locate, and those who helped you locate it would be criminals working outside the law. An "assisted-suicide machine" is more likely to be legal, and easier to locate; and those administering it would be respected health care professionals earning salaries and paying taxes.

There are two interrelated definitional lessons to take from this example. First, changes in definition emerge within real cultural struggles. Kevorkian's public flaunting of the use of his machines is clearly an attempt to force a legal and cultural change in what the machines mean and what matters. His efforts, and the lawsuits and debates that involve his efforts, may significantly affect the ways that people understand life and death. All meaning changes in struggles like this, although the struggles are not always as dramatic All meaning changes in struggles to *make something mean* in particular ways.

Second, the definitional move we propose—away from the equation of technology with "thingness" and toward a notion of technology as articulation and assemblage—clearly matters. The two Kevorkian "machines"—the "killing machine" and the "assisted-suicide machine"—are only the same machine if you think solely in terms of their "thingness," as discrete objects that exist apart from other objects and bodies. They are clearly different machines if you admit that what they "are" interpenetrates the lives, bodies, and objects of which they are a part, and that the forms of this interpenetration can differ. By understanding them as different machines, we are compelled to explore the culture, the cultural arrangement, and the flows within which these machines come to have a variety of meanings. We learn, as a result, more about everyday life, and more about technology as part of everyday life. Therefore, it is important to struggle with the problem of definitions and definitional change. That is, in part, the way the world changes.

CHAPTER NINE

Causality

Beyond Determinism

As DISCUSSED IN CHAPTER 3, technological determinism and cultural determinism represent two extreme positions with very few options for understanding how change happens. Most people, it turns out, think in more varied and often more complex ways about cultural and technological change. In her book *Communication Technologies and Society*, Jennifer described the most salient ways that people understand technology and change. She developed a way to explore the structures of thinking about causality used by people when they think about, make statements about, or take positions on technology.[1]

Jennifer proposed that thinking about technology falls into two major categories (or perspectives): *mechanistic* perspectives on causality, and *nonmechanistic* perspectives on causality. Within each of these categories there are sub-categories (or perspectives): The two mechanistic perspectives are *simple causality* and *symptomatic causality*. The two nonmechanistic perspectives are *expressive causality* and *articulation and assemblage*; and between them is a perspective that strives to be nonmechanistic, but doesn't quite succeed. This in-between position is called *soft determinism*. The grid looks like this:

Mechanistic Perspectives		Nonmechanistic Perspectives
Simple Causality		Expressive Causality
	Soft Determinism	
Symptomatic Causality		Articulation and Assemblage

As you will see, this grid incorporates technological determinism and cultural determinism, but it transforms them in a way that makes it possible to characterize the more complex ways people think about culture and technology in everyday life.

Mechanistic Perspectives on Causality

When people understand change from a mechanistic perspective on causality, they think and act with four basic assumptions. These assumptions are not necessarily held consciously, although they might be. Usually, however, it takes careful reflection (sometimes self-reflection) to see that the assumptions are at work. The four assumptions are as follows:

Assumption #1:
Technologies Are Isolatable Objects, That Is, Discrete Things

The idea or definition of technology that comes into play when someone takes a mechanistic perspective on causality is that technologies are objects, artifacts, and things. Recalling the discussion of definitions in the previous chapter, technology here means "thing," with the consequence that it draws our attention away from the context within which the artifacts are produced and used. Technology is thus isolatable, meaning that we assume we can examine the technology itself, without having to consider, as part of what it is, the people who develop and use it, or the culture within which it is developed and used.

Assumption #2:
Technologies Are Seen as the Cause of Change in Society

This assumption should now be familiar as a technological determinist position. When someone takes a mechanistic perspective on causality, discrete technological objects have effects on the culture and not the other way around.

Assumption #3:
Technologies Are Autonomous in Origin and Action

Autonomous means that something is separate, discrete, and independent. To say that technologies are autonomous is to say that they are discrete things that function independently. What they are and what they do does not depend on a relationship with anything else. To be autonomous in origin means that technologies come into being independently. To be autonomous in action means that technologies act and have effects independently.

What does it look like to come into being autonomously? There are three ways that people talk about technology that suggest they assume it has autonomous origins. First, people sometimes talk about technologies as though they simply *appear*: they materialize "out of thin air," and "drop from the sky." They are the ultimate *deus ex machina*, literally in Latin, "the god in the machine." This phrase refers to the practice in medieval theater of dropping a mechanical device with a "god" aboard it onto the stage. The god's function was to resolve the conflict of the drama with no other apparent connection to the story (or context). Thus, no matter what seemingly irresolvable turn the story might take, the deus ex machina arrives "out of thin air" to set all things almost magically straight. In a similar way, people often assume that technologies appear as though motivated by some inertial force that exists apart from the goals, motivations, and desires of human beings and apart from the organization of culture. They are dropped from above to resolve (or create) conflict.

An additional image captures the way this belief in the autonomous deus ex machina works. Imagine for a moment that culture is a pool table covered with balls at rest. A new technology (the cue ball) drops on to the table, appearing from outside the culture. Once dropped, the technology/cue ball collides into the other balls, creating change. The new technology, like the cue ball, is understood as though it comes from somewhere outside culture, with no preexisting relationship to the culture it affects.

Does the new technology simply appear, apart from the influence of culture? Most people would answer that it doesn't; they will add that someone had to invent it, build it, use it, and so on. But if you listen to what people say about new technologies, and if you watch how they interact with them, you will see that they do in fact often assume that technologies appear in this autonomous way. A typical newspaper article on computers might begin with a statement like, "Computers have changed education since they were introduced into the schools in the 1980s." The article then details various effects caused by the computer. But, we might ask, where did the computer come from? How was it developed and why? Who introduced it into the schools and why? How was it taken up and used in the schools? And, most important, in what ways do the answers to these questions offer insight into the effects that we observe? The impression that technologies arise autonomously is reinforced by the absence of such questions (and the absence of answers to these questions) in the discourse about technology.

The second way that the origins of technology are treated as autonomous occurs when people consider the beginning of the technology as though it were simply created in the imaginings of an inventor. Just as in a comic strip, where the convention used to indicate an idea is the dialogue balloon with a light bulb lighting up, new technologies are like lightbulbs in the mind of an inventor. They simply light up, go off, appear in a flash. They are autonomously born out of the air. In this case, the inventor, like the cue ball, is outside culture; the inventor is considered to be a unique being or a genius who simply comes up with ideas that appear like a flash.

But do such inventors and their inventions appear apart from the influence of culture? Most people would answer, "of course not." But, again, if you listen to what people say about technologies, and if you watch how they interact with technologies, you will see how pervasive is the belief in their autonomy. We learn this way of thinking about, and interacting with, technology in grade school, when we are taught, for example, that Eli Whitney invented the cotton gin in 1792, Robert Fulton invented the steamboat in 1802, Alexander Graham Bell invented the telephone in 1876, and so on. We learn to associate particular technologies with individual inventors rather than with a particular cultural context within which technological solutions are searched for, invented, and developed.

The practice of granting patents reinforces this way of understanding the process of invention. Patent practice only recognizes individuals as inventors. Patent seekers must prove to the satisfaction of the patent office that the invention is truly theirs and theirs alone. Even though patent rights may be assigned or sold to a company, corporation, or another individual, inventions are not understood to "belong" inherently to the culture within which the inventor lives and works. Rather, inventors have the right to prohibit others from using or producing the invention. When technologies are regarded and treated as unique acts of invention in the minds of isolated inventors, the culture reinforces the understanding that technologies arise autonomously, which reinforces, in turn, the privileging of the individual in culture generally.

The third way that technologies are understood as autonomous in origin is that technologies are sometimes seen as self-generating. In this way of thinking, technologies give birth to other technologies. Nobody, when pressed, would say that technologies actually give birth, yet it is common to hear people say that one invention gives rise to another: People make statements like, "the internal combustion engine gave rise to the automobile," and "the radio begat television." Without questioning what it means to "give rise to," we often talk about technologies as if they give birth, without the aid of any cultural influences, or even inventors. Thus, technologies are treated as though they simply arise autonomously.

If technologies are understood to arise autonomously in the above three ways, it follows logically that people would understand them to act autonomously as well. If the very appearance, or birth, of a technology, is free from cultural influence, it stands to reason that it does not need culture to do what it does. It acts independently, and its effects are the effects of the objects and artifacts, the stuff, the isolatable things, not the effects of cultural choices or arrangements. Further, because it acts autonomously, it acts with impunity, as an amoral force that cannot be held responsible for its effects, whether for good or evil. It simply exists and it simply acts. Members of the culture upon which it acts are virtually helpless in the face of this enormous, autonomous force.

Assumption #4: Culture Is Made Up of Autonomous Elements

Once you understand how technologies are seen to arise and operate autonomously, it is easier to understand how, when people utilize a mechanistic perspective, every aspect of the culture is seen as autonomous. The image of the pool table can serve again to illustrate. Think of culture as all those individual pool balls lying at various positions on the pool table. Each component of culture—economics, politics, law, religion, the family, education, music, and so on—is understood, like a pool ball, to be a separate phenomenon, each without any intrinsic relation to the others. Music, for example, would be understood to develop in a particular way totally unrelated to politics or law or the family, and so on. One might have a momentary effect on another—like when a pool ball strikes another—but the essence of each remains intact. Family and religious values may have an effect on the law, which may have an effect on a music rating system, which may have an effect on music. But the music is still music; it isn't in an intrinsic way about family values, religion, or the law. After the effect, music goes on being, in its own independent way.

This is important because, from a mechanistic perspective, the meaning, significance, and role of something such as music or technology is understood by focusing on the thing itself. To understand music, you would study music, not law. To study any component of culture, such as technology, you look at the thing itself even when you might have to acknowledge the momentary effects that other components of culture occasionally have on it and the effects it has on the other components of culture. It is as though culture is made up of all these independent entities sitting on the pool table waiting for technology (the cue ball) to come careening on the scene to put them all in motion.

These four assumptions form the backbone of a mechanistic perspective on technology. But if you closely examine the way that people make arguments from a mechanistic perspective, you will find that it takes two different forms: a simple causal form and a symptomatic causal form.

Both simple and symptomatic causality are mechanistic positions, and thus operate with the four assumptions discussed above. Where simple and symptomatic causality differ is in their understanding of the inevitability of effects. Simple causality assumes that *effects are inherent* in the technology and that *precise effects are inevitable*. Symptomatic causality assumes that *broad parameters of effects are inherent* in the technology, that *a range of effects is inevitable*, and that various social forces are responsible for steering, or choosing from among those effects.

Simple Causality

As stated above, when someone understands how technological change happens from a simple causal perspective, they assume that *effects are inherent in the technology and that precise effects are inevitable*. To say that the effects of technology are inherent in the technology implies that the effects are a natural and inseparable quality of the technology. To say that the precise effects are inevitable implies that once the technology appears it is absolutely certain that precise effects will follow. It is the nature of the technology that determines these precise effects. If someone believes in this way that the effects are inevitable, no force, no human, and no organization could shape or change them. The effects would therefore be unavoidable.

Recalling that the mechanistic perspective assumes that technologies arise and act autonomously, the simple causal perspective pictures technologies coming out of nowhere to exert uncontrollable and precise effects on culture, without any form of cultural assistance. We could only be passive recipients of these effects. We might choose to accept them, or, as some people put it, we can simply be left behind. We can be "on the bus" or "off the bus," but we can't do anything other than accept the fact that the bus will roll on down the road, with or without us.

When we look at the theoretical logic in this way, it's difficult to think that anyone could really believe in a simple causal perspective. However, when you examine the positions people take up in relation to technology, evidence abounds that a simple causal perspective is quite widespread. For example, when we have asked computer and engineering students why they have chosen these professions, they often respond with a simple causal argument. They argue that the computer, by its very nature and over which they have no control, is creating a world that determines where they will have to work if they want to thrive. The computer, in this answer, is singularly responsible for the changing nature of the workforce, as though it were an entirely autonomous force. Statements such as "it is inevitable that the computer will change the nature of employment" have become commonplace, and millions of people have made life choices based, at least partly, on that belief.

When looked at theoretically, simple causality is quite simply indefensible. To go back to the example of the gun introduced in Chapter 3: Everyone knows that guns don't materialize and kill people—someone has to pick one up. Remarkably, however, people do make these kinds of arguments. It is as though somewhere, deep down, many people believe that these technologies do have enormous, autonomous power to shape our lives all by themselves, and that there is absolutely nothing that anyone can do to alter those effects.

Symptomatic Causality

The symptomatic causal perspective probably represents the most commonly held position when people think about, talk about, and interact with technology. Though still a mechanistic perspective grounded in the four basic assumptions discussed above, it assumes a more sophisticated understanding of effects than the simple causal perspective. When people take a symptomatic perspective, they do not believe that *precise effects are inherent* in the technology and therefore exactly inevitable. Rather they believe that *a range of effects is inherent* in the technology and that there are choices that can be made within that inevitable range of options. For example, a simple causal argument might maintain that it is inevitable that guns will kill. However, from a symptomatic perspective, there are options open to us among an inevitable range of possible effects. Yes, guns will kill (that much is inevitable), but the range of possible effects might include: (1) killing game animals and not people, (2) killing game animals and people, (3) killing only criminals and not innocent people, and so on. In another example, a simple causal argument might maintain that computers will put people out of work. A symptomatic causal argument might agree that it is inherent in the computer to change the structure of jobs. However, the range of possible effects might include (1) increasing the number of unemployed people, (2), retraining those put out of work to take up new kinds of computer-related jobs, (3) retraining people to take up new kinds of non-computer related jobs. In this case, the range of options is inevitable; one of the possible effects will inevitably occur.

The difference between precise effects (in a simple causal perspective) and a range of effects (in a symptomatic causal perspective) is significant. While killing is an inevitable result of the gun (according to either perspective), only the symptomatic perspective assumes that there are any options to choose among regarding cultural responses to such killing. While the structure of jobs will change (according to either perspective), only the symptomatic perspective assumes that there are any options to choose among regarding the configuration of those jobs.

What then determines which effect within the range actually occurs? Remember those pool balls on the table? Like pool balls in motion, the variety of social forces (such as law, religion, economics, politics, family, etc.) may deflect the technology (the cue ball) so that one effect or the other occurs. So, for example, we might pass laws making killing other humans a mere misdemeanor, in effect encouraging people to use the gun to kill humans. We might develop a religious belief that renders it unthinkable to kill another human with a gun. In the first case, guns will kill game animals and humans; in the second case, the gun will kill only game animals. However, in both cases, the gun will inevitably kill. In the example of the computer and jobs, we might let those people who lose their jobs to computers fend for themselves. Alternatively, we might develop educational programs for retraining people to work with computers in new jobs or to take up new kinds of non-computer-related jobs. In the first case the effect will be unemployment; in the second and third cases the effects will be different kinds of reemployment. However, in all three cases, the inevitable effect is that the computer will change the structure of jobs.

When people understand change from a symptomatic causal perspective, they see that our choices involve more than simply adapting (or not) to technology, but to steering, directing, or choosing within the range of inevitable effects. The challenge is to figure out what the range of inevitable effects is, to evaluate those effects, and to develop creative ways to ensure that the better effect is the one that happens.

It is worth recalling, however, that the symptomatic causal perspective, like the simple causal perspective (because they are both mechanistic), does not assume that we can initially encourage or interfere with the appearance of the technology. Neither does it assume that we could avoid the inevitable effects. Technology is still assumed to appear autonomously, and it is still the technology rather than the culture that is assumed to cause the effects. From this perspective, there is nothing, or nobody, to blame or praise for its appearance (except, perhaps, for that genius inventor), or for the fact that it has certain inevitable effects. From this perspective, we do have some responsibility, for we are charged with shaping the outcome within the inevitable range. We are limited, however, for we can only steer to one side or the other as we careen down the road on which technologies inevitably take us.

Reaching for a Nonmechanistic Perspective: Soft Determinism

In response to the complexities of studying technology, scholars have come to resist thinking of technology as being either autonomous in origin or as the sole agent in causing effects. For example, a workshop at MIT on the question of determinism was held for a group of such scholars in 1989. Their discussions resulted in a provocative book that explores the problem, *Does Technology Drive History? The Dilemma of Technological Determinism*. The introduction to the book, by Leo Marx and Merritt Roe Smith, proposes a two-stage causal process called "soft determinism."[2] The soft determinists recognize that "the history of technology is a history of human actions," implying that every technology has an origin in human actions.[3] The task of the soft determinist is to describe the particular action, or *critical factor* that gives rise to a technology to begin with. For the soft determinist, the critical factor is *the original causal agent in a chain of causality*. For example, the irrepressible human desire to create may be seen as the critical factor, or initial cause, of inventing the gun. After the gun is invented, though, it takes on a life of its own and has effects on its own. Thus, even if it had its origins in human actions and is not autonomous in origin, the technology still acts autonomously.

Soft determinism thus acknowledges the importance of the cultural context within which a technology originates. It tends to remain, however, a form of determinism, like simple and symptomatic causality, because the technology, with "a life of its own," acts by itself to generate effects. Although it is a significant attempt to resist the problems of mechanistic causality, soft determinism simply extends the cause-effect relationship back to particular critical factors (such as economic, demographic, intellectual, and cultural factors) that act as a prior cause.

In response to the soft determinists, we might ask: How can a technology's actions be autonomous if its origins are not? Don't we deploy technologies in particular ways, steering their effects to some degree? Wouldn't this imply partnership in the generation of effects? This is, after all, the implication of symptomatic causality: that effects can at the very least be steered. Marx and Smith realize that as soft determinists sort through the various social, economic, political, and cultural causes, they often end up describing a complex matrix within which technologies originate. It becomes difficult to identify a single critical factor in a simple causal chain in this situation. As Marx and Smith observe, causal agency becomes so deeply embedded in the larger social structure "as to divest technology of its presumed power as an independent agent initiating change."[4] If the origin of a technology is so caught up in a complex cultural matrix, how can it be said to have effects independently of this matrix? If, for example, the gun is developed because we are a hostile species inclined to kill those we perceive as enemies, *and* because we have a pressing need to kill a particular enemy, *and* because we have already developed gun powder, *and* because we revel in the intellectual challenge of invention, *and* because we have a religious sanction to kill, and so on, how then can we say that the gun causes killing, rather than the relationship between the gun and culture? This observation suggests that technology does not act autonomously. Consequently, if neither the origins nor the actions of technologies are autonomous, we unseat technology from its role as the central defining causal agent in cultural change. Technological determinism and its various forms—simple causality, symptomatic causality, and soft determinism—are insufficient to explain the role of technology in culture. Instead, we need to know more about the matrix within which technologies are developed and used. We need a better way to understand the complex process within which there are effects. This is what the nonmechanistic perspectives attempt to accomplish.

Nonmechanistic Perspectives of Causality

When people understand change from a nonmechanistic perspective of causality, they think and act with three basic assumptions that differ dramatically from the mechanistic assumptions. Again, these assumptions will not necessarily be held consciously, although they might be. Again, it often takes careful reflection (sometimes self-reflection) to see that the assumptions are at work. The three basic assumptions are as follows.

Assumption #1: Technology Is Not Autonomous, but Is Integrally Connected to the Context within Which It Is Developed and Used

When people assume that technology is not autonomous, they assume that it is not a discrete isolatable thing. This is where definitions, as we discussed in the previous chapter, begin to matter enormously. If people define technology as integrally connected to the context within which it arises, it cannot (by definition) come from outside the culture. It cannot drop from the sky or simply appear like a bolt out of the blue. It cannot be understood to be like a cue ball introduced onto the pool table from outside. Technology, if it is not autonomous, is always a complex set of connections, or relationships, within a particular culture—not a thing, but a structure of connections. Within these connections, things emerge and are used, but the "thingness" of a technology is only one aspect of what it is. The rest of what it is can only be understood by describing the nature of the connections within which it is developed and used. For example, rather than thinking of the gun as just a material object, it might be understood as a thing developed, designed, and used to kill enemies. That *is* what it is; it is the connection between the thing, the desire to kill, and the practice of killing enemies.

Assumption #2: Culture Is Made Up of Connections

When people assume a nonmechanistic perspective, not only is technology understood to be a structure of connections; culture is also understood to be a structure of connections. In this way of thinking, culture is not just a bunch of unrelated components that are scattered like pool balls on the table. Rather, culture is a complex set of connections or relationships: more like the formation of the balls on the table than the balls themselves. No particular cultural component, such as education or the economy, stands alone. What they are is the set of connections or relationships among forces. For example, if you wanted to understand education from this perspective, you must understand its connection to the economy, for education is integrally bound up with economic developments. You must understand its connection to technology, for education is integrally bound up with the role of technology in relation to the economy. A rich understanding of education would require understanding many more connections among the cultural forces that animate the practice we call education. From this perspective, then, *culture is the constantly changing web constituted by these connections.* Every phenomenon in the culture (including technology) would have to be understood as part of that complex web.

Assumption #3: Technologies Arise within These Connections as Part of Them and as Effective within Them

People who take a nonmechanistic perspective do not regard technology as being either a simple causal agent or a simple effect. In fact, in a nonmechanistic perspective, the language of cause and effect no longer suffices. Rather, adherents to a nonmechanistic perspective draw attention to the ways that technologies emerge in shifting connections of forces and are part of those connections and forces. In this view, technologies emerge from within a context, as part of that context, and in relationships to forces that have effects. The challenge for nonmechanistic thinking is to explain this complex process of change.

These three assumptions form the backbone of a nonmechanistic perspective on technology. But if you closely examine the way that people make arguments from a nonmechanistic perspective, you will find that it takes at least two different forms with respect to understanding how the cultural web is put together: an expressive causal form, and a form we call "articulation and assemblage." These two perspectives differ principally in the way they understand how the cultural web is put together. When someone takes an expressive perspective, he or she believes that one force or connection takes center stage and gives a uniform shape to a homogeneous whole. When someone takes an articulation perspective, he or she believes that no force or relationship takes center stage, and that the whole is more heterogeneous (though a heterogeneous whole sounds counterintuitive; the concept will be explained below). As a result, adherents to each of these perspectives understand change to occur somewhat differently. While adherents of each perspective will explain the emergence, development, and use of technologies as things as well as connections, each group conceives the process differently.

Expressive Causality

When people understand culture and technology from an expressive causal perspective, they understand cultural connections structured by one essential element, very much like the soft determinist's "critical factor." However, where for the soft determinist there might be different critical factors in relation to the development of different technologies, for the expressive thinker there can only be one critical factor for all of culture in every instance. That means that every human action, every cultural force, every connection, and every thing is given form by the same element. The term for this element is "essence." And just as some people believe that an individual human being has an essence, such as being essentially "good," the expressive causal thinker understands culture, and every aspect of it, as having a single and shared essence.

What then is the essence of culture? That depends. In the same way that soft determinists might disagree about the critical factor responsible for producing the gun, expressive thinkers might disagree about what the essence of culture is. For some it is inventiveness, capitalism, aggression, or, in the case of the most famous expressive thinker, Jacques Ellul, it might be something called "technique."

Ellul and his concept of technique provide an excellent example of how an expressive causal thinker sees culture and technology. Technique, for Ellul, is "the totality of methods rationally arrived at and having absolute efficiency (for a given stage of development) in every field of human activity."[5] Technique is thus the application of rationality and efficiency. Ellul insists that technique is the essence of modern culture. He writes, "Technique is not an isolated fact in society...but is related to every other factor in the life of modern man."[6] Thus, to understand technology, we must understand technique. To understand the family, religion, politics, economics, or anything else, we must understand technique. Because technique permeates and defines everything, every connection or cultural manifestation—such as the relationship between the gun and politics—can be explained in terms of technique. Not all expressive thinkers posit technique as their essence, but whatever essence they posit would be the essential element that explained all else in culture.

If for expressive thinkers all connections and manifestations can be explained in terms of an essence, how do they imagine that change occurs? For them, an essence evolves. Like a rosebud opening into a rose, an essence, while always there, unfolds and becomes more fully what it is. As the essence unfolds, all of its expressions throughout the culture evolve along with it. They *reflect the essence*, they *manifest the essence*, and they *enhance the essence*. Thus, the whole culture, what is called a "cultural totality," evolves together. The whole of culture takes on a strikingly homogeneous form in this way of understanding.

Technologies, from an expressive perspective, are understood to reflect the essence, manifest the essence, and enhance the essence—whatever the particular essence might be understood to be. If the essence is seen to be the Ellulian technique, then any and every technology has technique as its essence. Every technology reflects technique, manifests technique, and enhances technique.

Since a cultural totality is understood to have only a single essence and that essence is the critical factor defining everything in that culture, nothing can escape being defined in those terms. It is not be possible to create a technology that does not share that same essence. Thus, there can be no fundamental contradiction with the essence. Technologies can only be of one evolving type, and nobody is free to change that. The gun, for example, can only be a reflection, manifestation, and enhancement of an essence. If the essence is technique, then the gun is a rational and efficient method. That it kills is almost not the point. The very fact that we kill is not an effect of the gun, but a reflection, manifestation, and enhancement of a rational and efficient means of interacting. What more efficient way to rid yourself of enemies than to kill them? The significance of the gun is that it kills more rationally and efficiently than weapons that precede it. That is the point. Further, there really is nothing that we can do about this situation; every facet of life is caught up in the inexorable march of our cultural totality toward increasing rationality and efficiency. We can only stand by and observe, perhaps waiting for a catastrophe so large, or a god so kind, as to intervene so radically that the essence changes.[7]

If all of this sounds a little crazy and you are thinking that nobody but some philosopher (which Ellul was) could come up with something like expressive causality, stop and observe a bit more carefully. The Unabomber's fatalism, discussed in Chapter 7, can be attributed to his understanding the industrial-technological system in expressive causal terms. But there are also plenty of examples of a less dramatic nature. If someone protests that the telephone, television, or computer systems are developed only to make money, the response is often to point out that in a capitalist system "that's just the way it is." All new communication technologies, it is often argued, must be developed as capitalist enterprises—they wouldn't be viable otherwise. That is an expressive causal position. If the essence is capitalism, then every aspect of culture can only reflect, manifest, and enhance capitalism. The only difference is that the underlying assumptions that accompany the argument are typically left unstated, unexplored, and unchallenged. As with all the perspectives considered thus far in this chapter, they are invoked in day-to-day language and practice, but the assumptions on which they depend are rarely examined.

Articulation and Assemblage

The concepts of articulation and assemblage, as they have developed in cultural studies, provide an alternative to the perspectives on causality presented above. Because articulation and assemblage are so central to understanding the orientation of this book and the direction we propose, we discuss them more fully in a separate chapter: Chapter 11. But we take a little space here to point toward the direction we are moving.

To think about technology as articulation and assemblage is to adopt a non-mechanistic perspective, and thus operate with the three nonmechanistic assumptions discussed above. Articulation and assemblage assume: that technology is not autonomous, but is integrally connected to the context within which it is developed and used; that culture is made up of such connections; and that technologies arise within these connections as part of them and as effective within them.

As such, articulation and assemblage share crucial features with an expressive perspective. However, articulation differs from expressive causality in significant ways. First, while it does hold that culture is made up of connections, it does not insist that all these connections are reducible to an essence or to a critical factor. Instead, culture is understood as being made up of myriad *articulations* (intermingling elements, connections, relationships) that make some things possible, others not. These articulations are sometimes corresponding, as they would be in an expressive perspective, where they share a common form; but they are also sometimes noncorresponding or even contradictory. Articulations are dynamic inter-minglings that can move in many and various directions, propelled by various and changing circumstances (of other articulations). The "web" of these articulations is what we call an *assemblage*.

Within a particular assemblage, technologies are developed, used, and have effects. In so doing, new articulations are constituted in a revised (or rearticulated) assemblage. As philosophers Gilles Deleuze and Félix Guattari put it, technolo-gies exist "only in relation to the interminglings they make possible or that make them possible."[8] Because technologies only exist in relation to these articulations, they are themselves articulations. Technologies come into being, are developed, are used in the dynamic movement of an assemblage, and they are diffused within the assemblage. They *are* assemblage, in that they are made up of webs of cor-responding, noncorresponding, and contradictory articulations. Therefore, no technology has one single essence, definition, purpose, role, or effect.

Thinking of technology as articulation and assemblage allows us to take se-riously the implications of Eddie Izzard's playful insights about guns raised in Chapter 3, and apply these insights to any and all technologies. We no longer need to decide if guns kill people or if people kill people, because we no longer see the problem as attributing causal power or responsibility to one or the other—to the technology or to the culture. Instead, the relevant task, when utilizing this perspective, is to map and critique the assemblage (what we have previously called context) within which different uses and effects are both possible and effective. A complex cultural assemblage produces technologies (such as guns) as particular, contingent kinds of tools to be used in particular, contingent ways. Similarly, a complex cultural assemblage takes up technologies (such as guns) and uses them in particular, contingent ways with particular, contingent effects. Because an assem-blage is made up of multiple (corresponding, noncorresponding, and contradic-tory) articulations, change takes place in the dynamic tensions among the articula-tions that constitute an assemblage.

Clearly, this perspective is nonmechanistic. But what you have just read is, admittedly, a little difficult to "unpack." The new vocabulary you need, the new concepts to work with—articulation, assemblage, and contingency—are explored in much more detail in Chapter 11. Regardless, it will be helpful to explore the concept of agency first, which we undertake in Chapter 10.

Conclusion

It is important to remember than anyone who thinks or writes about technologies, anyone who makes a decision involving technologies, and anyone who interacts with technologies, lives out an understanding of one or some combination of the above perspectives technology: simple, symptomatic, expressive, and articulation/assemblage. That is clearly true for all of us. Whether we think these matters through theoretically or not, we internalize a scheme of how technology works and what role it plays in our lives. Throughout the many years of listening to what people say about technology and watching them live out a relationship to it, we can say without hesitation that most people tend to be inconsistent in their understanding of what technology is and how it works. For example, a person might be against gun control because people, not guns, kill people (a symptomatic causal perspective); but they might be opposed to computerized banking because the machine depersonalizes banking (a simple causal perspective). This inconsistent thinking points to the likelihood that other cultural forces and connections (beyond the purely theoretical) come into play, that is, articulate, in the decision-making process. By thinking through the problem from the perspective of articulation and assemblage, we can begin to see the power with which some of these other forces and connections shape technological culture, our understanding of it, and, finally, our responses to it.

CHAPTER TEN

Agency

From Causality to Agency

THERE IS A COMPUTER SITTING ON GREG'S DESK. This is hardly a surprising disclosure in this day and age. Actually, if we want to be accurate, there's a computer monitor sitting on his desk; there's a mouse and a mouse pad to the right of the monitor; there's a keyboard mounted on a nifty sliding drawer just under the desktop; and the computer sits on the floor under his desk (unfortunately, just within range of his idly kicking foot).

We begin with such a banal example because instances from everyday life allow us to address more easily the weaknesses of the received view of technology. For example, the received view of technological determinism would look at the scene described above and consider the computer to be the center of attention. It would see how the computer affects life, changes work habits, communication patterns, posture, and so forth. But this view ignores much of what else is going on: it ignores the yellow sticky notes attached to the monitor frame and screen, the orientation of the monitor to the door and window, the piles of papers blocking the mouse, the nature of the work being done, and so on. If it does notice these things, it sees them only as evidence of the effects of the computer on the way Greg works.

In contrast, the received view of cultural determinism would look at the scene in Greg's office and focus initially on Greg's activities rather than the computer. It would see how the computer in general and this computer in particular have been developed in response to the needs of computer users. The computer itself almost disappears from the picture, obscured by the functions for which it was developed and to which it is put. Here, the yellow sticky notes would be taken to represent some of those functions.

What is problematic with both these views is that in restricting their view of the office in their particular ways, they are unable to grasp or even recognize the articulation of broader cultural forces at work. Despite their differences, both positions view this situation through the same lens: that of causality. They are restricted to asking questions only about one dimension of the situation. They can only ask, on the one hand, what does the computer cause to happen in Greg's life? Or, on the other hand, how is the computer a response to the cultural wants and needs of people like Greg?

The causal approach has a certain universal undertone to it, meaning that its purported causal effects are assumed to be the same under any—and every—circumstance. The causal approach cannot adequately grasp the particularities of situations. It cannot, for example, differentiate the office environments of the co-authors of this book, both of whom work with computers. The causal approach talks about the effects of *the computer*, but is less helpful in discussing the effectiveness of *this* or *that* computer.

To obtain a richer view of what is happening in a particular situation, we propose a multidimensional view that acknowledges the work of articulation. A causal approach is too reductive to provide that acknowledgment. What makes this approach richer is that it is sensitive to the contingent interplay of a wider variety of factors. To insist that the interplay is *contingent* is to recognize that culture (or technology) is not a set of stable, unchanging, and fixed categories or components, but rather a set of dynamic, changing, and interrelated connections or relations. One way to gain access to the nature of these relations is to utilize the concept of *agency*.

What we mean by agency differs somewhat from the definition that is found in the dictionary. According to *Webster's*, agency is

> **1** active force; action; power **2** that by which something is done; means; instrumentality **3** the business of any person, firm, etc. empowered to act for another **4** the business office or district of such a person, firm, etc. **5** an administrative division of government with specific functions **6** an organization that offers a particular kind of assistance [a social *agency*].

The emphasis in this definition, consistent with popular usage, is that agency is the power and ability to do something, and it assumes an *agent* that possesses that power. An agent, according to the same dictionary, is "generally, a person or thing that acts or is capable of acting, or…one who or that which acts, or is empowered to act, for another [the company's agent]" to bring about a certain result.[1] What is important about the dictionary meanings and popular usage of agent and agency is that they are ultimately defined in terms of the human realm and assume *intent* behind every action. For example, if your intention is to communicate with your mother, you can either send a friend over to her house to tell her something, write her a letter and drop it in the post and have the mail carrier deliver it, telephone her, e-mail her, or walk over there yourself. The friend, the mail carrier, the letter, the telephone, the e-mail system, and even your own body can be called agents in this situation because each represents a possible means of achieving the original human intention. They are all intermediaries through which you exercise your agency. Agency is thus something you possess. It is measurable in the sense that you can have more or less of it. Much of the world does not have access to many of these agents; thus their agency, their ability to secure a particular effect, is limited. As these examples illustrate, the dictionary definition and popular usage reduces agency to a thing, a possession of an agent, rather than recognizing agency as a process or a relationship.

We propose two changes to this view of agency. First, agency does not require human intention, which means that technologies can also be involved in relations of agency. Second, agency is not a possession of agents; it is a process and a relationship. The remainder of this chapter will set out each of these changes in turn.

Technological Agency

First, in response to the assumption that agency ultimately resides in human intention, we propose that technologies are active participants in everyday life, and can be seen as participating in relations of agency. Even though you are talking with someone on the telephone, isn't the phone itself part of that conversation? We tend to ignore the phone because we think of the conversation in terms of its content (what is said). But just as the tenor of a conversation changes depending on the individuals involved, the tenor of the conversation changes depending on the technology involved. For example, you might have to shout because of static or a weak signal. You may have to be thoughtful about talking in turn because you are using a cell phone. You might talk quickly because you are paying per minute of use. You might be able to walk around because the phone is cordless, and so on. The shape of the conversation in these cases cannot be reduced to simple human intentions. The technology matters.

When we ignore the technologies, we typically treat them as intermediaries, conduits through which intention, power, or action are achieved. However, as Bruno Latour has argued, the technologies are actually mediators, not intermediaries.[2] A mediator of a dispute is a person who steps between the parties involved and actively tries to get both sides to agree, to influence them in some way. A mediator is active and presumes a transformation: The demands of both sides in the dispute are altered to reach common ground. So in our example of a telephone conversation, the telephone (including both phones, their wires, and all the wires and switches and transmissions in between) is a mediator; it is one of the factors transforming the conversation.

In another example, backaches are often related to the posture you take when you work at a computer. In this case, it doesn't make sense to talk about either the user or the technology as consciously "intending" to give you a backache. Yet, the computer clearly plays a role in your backache. The technology adds something more than, apart from, or different from human intent. This is why it is incorrect to talk about technology as a mere "tool." It does something more, beyond, and apart from its intended "use." For this reason, technology theorists have developed the argument that technologies are also actors. This perspective is sometimes called Actor-Network Theory, and involves the concepts of *actors* (another term for agent), *translation*, *delegation*, and *prescription*, each of which we will discuss below. Actor-Network Theory is a useful approach for thinking about agency, a useful place to begin. However, there are some problems with the way the approach has developed. So, we first explain the position and then attend to the problems in order to move beyond them.

Actors

French sociologists Michel Callon and Bruno Latour define an actor as, "Any element which bends space around itself, makes other elements dependent upon itself and translates their will into a language of its own."[3] Let's break this definition down and explore each part. First, "any element which bends space around itself." What does it mean to bend space? Imagine that you see a strange dog snarling on the sidewalk. You might respond by slowing down, changing your path of travel, your attitude, and your behavior. That dog has altered, or bent, the space around it. Likewise, a computer shapes the space around it. While working at the computer, you assume a particular posture, even an attitude. You arrange the elements of your desk or table, place the desk or table close to an outlet, a phone jack, and so on. You might wear particular glasses or hold your hands in a particular way. That computer has altered the space around it, bent space, and bent you as part of that space.

Second, it "makes other elements dependent upon itself." A technology is never alone or isolated; it is always connected with other actors, that is, with other technologies and people. Any network of actors consists of an indeterminate number of relations of dependence and control. For example, people become dependent on computers in many ways, such as to communicate with others via e-mail or instant messages, to check spelling and grammar, and to calculate math functions, such as balancing a checkbook. Likewise, the computer is dependent on other actors in many ways, such as for repair, programming, start up, electricity, and general implementation. Technologies make other elements dependent upon them. Technologies also depend on other elements.

Third, it "translates their will into a language of its own." Translation implies an altering of form. In terms of actor-networks, to translate means *to alter the form of something to bring it into alignment with the technology, system, or culture*. For example, we translate a sentence from one language to another to facilitate understanding. Technologies translate crude oil into a form so that other machines can use it. Computers translate human language into machine language so that the computer can process it. When you write a letter on a computer you translate your thoughts into a particular posture as you sit at the keyboard and enact the particular movements of typing. When you take a multiple-choice exam, you translate the answers in your head into an appropriately filled bubble on the page. Translation is the process of transformation. The function of a mediator is to translate and transform. An actor—whether human or technology—is a mediator.

Delegation

Latour gives another name to the process of translation: *delegation*. (Since this is quite a different way of talking about technology, theorists like to try out a number of different metaphors or terms to try to grasp just what it is they are getting at.) To be a delegate means that you are representing someone else (or many people), and speaking and acting (for example, voting) on their behalf, like a representative in a democracy. Delegates speak on our behalf to political conventions, international bodies such as the United Nations, or in negotiations. To delegate means to hand over a task or tasks to someone (or some*thing*) else. We delegate tasks to humans and nonhumans, such as technologies.

When humans delegate tasks to technologies, the technology does something a human used to do (direct traffic, open doors, assemble cars) or performs a task that humans could not do but wished they could (fly). Let's take the example of a bread machine. This is a machine to which a variety of tasks has been delegated. Tired of mixing ingredients, kneading dough, baking, and so on, humans invented a machine to do all this. All the human actor has to do is measure and pour in the ingredients, shut the lid, plug it in, turn it on, and the machine does the rest. So in this way a tiresome task has been delegated to a machine. We have translated much of the human work of making bread into a machine process.

Latour gives the example of a door. In his revelation, a door is a technology that makes walls more convenient. Walls are wonderful at keeping things in (warm air, small children, prisoners) or out (wind, bugs, barbarians), but they also keep us in or out. If we need to leave a room, we need a hole in the wall, which defeats the whole purpose of the wall: Now whatever is outside can come in (and vice versa). Latour argues that we take all the work of tearing a hole in the wall, climbing through, and rebuilding it again and translate that work into the work of opening and closing a door, which is much easier, much less time consuming, and much less messy. It is similar to our bread machine, which takes all the tasks of making bread "by hand" and translates them into the tasks of making bread "by machine" through an act of delegation. At the same time, bread has been translated into a "language" the machine understands. The machine does not make just any or every bread, but only a specific type of yeast-based loaf bread. It is a culturally specific machine performing a culturally specific task to make a culturally specific loaf of bread.

We don't just translate or delegate tasks to technologies—although this tends to be the general direction of delegation in our culture, since we believe that technologies are more reliable or stable than people. Tasks get translated (or *inscribed* or *incorporated*) into humans as well. For example, if we know how to drive a car, certain habits and skills are inscribed in our bodies, as are the rules, regulations, customs of the road, and accepted practices of negotiating traffic. We certainly don't need to be reminded of them at every turn and rarely even think about them, but we do perform them. Latour uses the concept of "prescription" to explore the effects of this kind of delegation from technologies back to humans.

Prescription

To continue with the bread machine scenario: Once the machine has been inscribed with tasks and is released into the culture, or at least placed on our kitchen counters, it prescribes back to us what bread *is*, as well as an enormous range of behaviors, attitudes, and values. Latour refers to prescription as "whatever a scene presupposes from its transcribed actors and authors."[4] Some of these presupposed behaviors, attitudes, and values have to do with bread. The machine presupposes a desire for a quantity of a particular type of bread, the availability of particular ingredients, and a particular, narrowed, or shifted taste. In other words, the machine translates tastes into a form consistent with its function. It is quite a taskmaster as well, demanding exactitude in ingredients and proportions of ingredients, or else it will not produce good bread. It requires cleaning and it requires that its owner find a space for it somewhere. In other words, it presupposes that you will be exacting in following directions, in maintaining and cleaning the machine; and it presupposes that you have space, ingredients, and a power source. It also presupposes that it was put together competently in the factory, that it was programmed correctly, that the delivery person did not drop it on the way to the store, and so on. The machine also prescribes behaviors, attitudes, and values having to do with technology, such as reinforcing the valuation of convenience and efficiency. It prescribes expectations for the proper household; if you have the machine you now are expected to produce fresh bread daily and eat it. In a sense, the machine demands that you make bread regularly to justify the machine.

The cell phone offers another example of the way that a technology prescribes behaviors, attitudes, and values back onto humans. When a person wanted to use the phone before the invention of cell phones, they had to get to a telephone. Now the cell phone does the traveling for us. So, in a sense, people delegated the task of traveling to the phone to the cell phone itself. But the cell phone prescribes back a daunting range of behaviors, attitudes, and values. First it demands that a person carry it; if you don't carry it, you can't use it. Beyond simply carrying it, a person has to buy batteries for it, pay for cell-phone service, and pay regular monthly bills. The prescriptive work extends still further. Now a person is expected to use the cell phone in places where there had been no telephone before: in restaurants, in automobiles, on vacations, while mountain climbing, and so on. People, when they wished, used to be able to be out of phone contact, but there is barely a place where that is possible anymore. The prescriptive pressure is to always be in contact. Thus, it becomes a good thing—if not a necessity—to have a cell phone while mountain climbing or in an automobile, because a climber or motorist can call for help if need be. Thus, the cell phone prescribes the value of always being in contact, of always being "on call," and works at obliterating privacy and the idea that privacy might be desirable. A whole new standard of expectations about being available is emerging as the cell phone (and e-mail) gradually blankets the planet.

Network

Actors are mediators, involved in processes of delegation, which include processes of inscription and prescription. So then what do we mean by *network* in Actor-Network? A network is a "summing up" of the relations between these actors and all these processes.[5] Networks are maps of *articulations*, which at this point you might think of as connections. The task of Actor-Network Theory is to discover how such networks get built, how they are maintained and transformed, how the articulations are made and unmade, and what qualities comprise those articulations.

We are being pretty abstract here; so we will give you an example. Let's talk about making bread by hand, before those bread machines became so trendy. You cannot make bread on your own, because it does not spring, fully baked, from your forehead! You need to gather the ingredients (eggs, flour, water, yeast), which you will articulate (connect) in a certain way to make bread. You also need to enlist the aid of other actors: a bowl or two, a rolling pin, a countertop, an oven, and so on. You are building a network right there in your kitchen. However, it does not stop there: For the eggs you need to enlist a chicken, which might mean walking next door to the barn and disturbing the chickens. Even if there are chickens waiting, you had to previously enlist a barn, chicken feed, and so on. For the flour, you might need to enlist the help of your pickup truck to get to the store to purchase it. The store didn't make the flour, so you need to follow the network further to include the distributor, manufacturer, milling machines, granaries, farms, and so on. And you haven't even started kneading, rolling, patting, or baking yet!

Here is another banal example: Last night Greg was heading out to teach his graduate seminar on technology; the topic was Actor-Network Theory, believe it or not. His hands were full, with a plastic-wrapped tuna sandwich on a plate, hot coffee, and books; and he found himself faced with a closed door. In order to get through the door, he enlisted the aid of a passing student, who kindly held it open for him. That is an obviously contingent articulation: He can't assume that this student will always be there. It worked once; it might not work again. If we wanted to stabilize this articulation, it might be more reliable to delegate this task to a nonhuman. In this case, we could delegate the task to an electric door opener activated by a button near the door. This is a more stable network, although it too can still break down.

While we have chosen to explain Actor-Network Theory using rather mundane examples (to make the process obvious), it is possible, using Actor-Network Theory, to talk about any situation (the Congress, a hole in the ozone layer, electric cars) with the same terminology. In every case, each moment of enlisting is a process of delegation that prescribes back a range of requirements. The resulting networks are more or less stable: The network of production and distribution of Pillsbury flour may be more stable and reliable than the network of production and distribution of eggs to a local farmer's market. An electric eye may be more stable and reliable than a passing student. To make a network more stable, we usually add more nonhumans to the mix. But with that stability and reliability come a plethora of prescriptions with which we must operate. The distribution of responsibility has merely shifted, albeit in significant ways.

Issues with Actor-Network Theory

You may have noticed that in our discussion of Actor-Network Theory we seem to have slipped into using a construction we initially objected to: referring to technologies as objects possessing agency. We did this in constructions such as "the cell phone prescribes," which suggests that the cell phone possesses the ability to make people respond in a certain way. We have done this in order to make it clear that what technologies do is not that different from what humans do. Technologies are not mere tools to be used, but active forces in the world. In saying this, however, we could be accused of anthropomorphism, acting as if machines have a will of their own, which is considered a "bad thing" if you are studying technology. Actually we *are* adopting a form of anthropomorphism here, but we don't see that as a bad thing. In popular discourse, we think of anthropomorphism as referring to a dancing tea kettle in a Disney film: The tea kettle acts like a human; it has a face; it sings; it dances. But as Latour points out, anthropomorphism means "either what has human shape or what gives shape to humans." So the cell phone or bread machine is anthropomorphic because (a) "it has been made by men" [sic]; (b) "it substitutes for the actions of people"; and (c) "it shapes human activities by prescribing."[6]

The danger here is less that of falling into a Disney-like version of anthropomorphism than it is the risk of restricting the attribution of agency to technologies alone and ignoring the activity of the network. The danger of thinking of technologies (and humans, for that matter) as agents in a network is that we then tend to think of actors as points in a stable web, like knots in a fishing net. This, Latour points out in his later writing on Actor-Network Theory, leads us into the misguided practice of separating the agent from the structure. Rather, the agent is the structure (the network) and the structure is the agent. And neither are stable things; although some versions of Actor-Network Theory have treated them as such. They are, instead, ongoing processes of delegation. In fact, Latour came to dislike the very term actor-network, because of the tendency for people to use it to separate agent and structure.[7]

The process of delegation does not just occur once, when the object is invented or manufactured, but over and over. When describing an actor-network or a map of articulations, we do not see a stable schematic before us, such as a map of the city or a diagram of a process, with all the elements and lines neatly and permanently set out. Instead, what we see is a series of constant movements, transformations, and circulations. We map brain to arm, to hand, to keyboard, to processor, to display, to server, to Internet, to something called the economy, and so on. We map a small packet of bread yeast to a store, to a distributor, to a manufacturer, to a bank, to a paper mill, to something called the economy, to something called politics, and so on. Each connection "to" is a delegation—Latour says it's like passing a ball in a sport.[8] Each delegation, which is a process and not an event, is a transformation. When you enlist something, you transform it. When the stove enlists electricity to bake bread, it transforms electricity into heat. The grocer transformed a pack of yeast into a profit. Greg changed a student into a door opener.

An additional, significant issue with Actor-Network Theory is that it tends to treat agency as if it were somehow universally available. The approach does not foreground variations in the availability of agency or the role of power in the construction and stabilization of networks. Simply put: Agency and power are not distributed equally throughout networks. The concept of articulation, with its sense of "lines of tendential force," is better at accounting for the unequal distribution of agency and power in networks. We turn to these concepts in the following chapter.

Conclusion: Why Agency?

What is the benefit of thinking of technology in terms of agency? For one thing, it greatly enriches and complicates our view of the world we live in. When we think of ourselves as moving through everyday life, we tend to focus on our encounters with other people and how these encounters alter the character of our day, or even our actions and behavior. What if we add to those encounters all of our encounters with (and uses of) technologies? How do our interactions with technologies contribute to the shape of everyday life? How does the process of delegation, inscription, and prescription account for what we do, think, and feel? How do technologies reinforce or give shape to rules and values from the mundane (when to cross a street) to the extraordinary (how to make war)? How do we choose to delegate to the technologies that we use (how to customize our computer desktop)? Even when a task has been delegated to a technology, that technology may still be enlisted to perform other tasks. You can use a technology for uses that it wasn't designed for or that its creators never dreamed of. You can use your bread machine as a mixer or a doorstop. You can "scratch" a record to make rap music. However, a technology is never completely pliable to your will, as you are always engaging a network of relations within which you are trying to rearticulate some of the connections while maintaining others. You can't get the sound of scratched human voices out of a bread machine (at least nobody has yet).

Our point here is this: If we continue to ask the question of which affects the other more (technology or people), we end up in a sort of philosophical tennis match (they influence us, but we influence them) that gets us nowhere. A more useful approach, we are suggesting, is to *ignore traditional questions about the division between technology and humans, and concentrate on analyzing the cultural field in which we live as a field of forces, relations, processes, and effects.* When we quibble about the origins of these effects, we often ignore the consequences, the real ways that life changes: how practices change, how values and beliefs shift, how power is distributed, how responsibility is transfigured. Now, those are issues that matter! In order to take these issues and this approach seriously, we have to look more closely at the work of articulation, a concept we have been developing and using indirectly throughout this book. Now we turn to look at it explicitly.

CHAPTER ELEVEN

Articulation and Assemblage

IF YOU WALK THROUGH TIMES SQUARE in New York City, your activities will be recorded by some of the 2,400 surveillance cameras concentrated in Manhattan. In 2002 the average New Yorker was photographed about 75 times a day.[1] Given that the number of cameras has been increasing in recent years, by now this number may be a modest appraisal of the real figure. Some are police cameras, some are traffic cameras, and some are private cameras installed by businesses and individuals for a variety of reasons. There are even cameras connected to the World Wide Web, so people can watch you on the Internet as you walk down the street. At a particular point you can even look up at a large electronic billboard and watch yourself walking.

The case of New York City may be dramatic, but it is hardly unique. There are surveillance cameras, microphones, and tracking technologies operating nearly everywhere. The average Londoner is photographed 300–500 times per day.[2] There are computer software programs that routinely track your movements from site to site on the Internet and monitor your purchases. Credit bureaus routinely gather and disseminate information on your borrowing behavior. With just your phone number, anyone with access to the Internet can get a map to your home or a satellite photo of your neighborhood. During the economic downturn, post-September 11, 2001, one of the few growth industries was that of surveillance technologies.[3]

The astonishing increase in the pervasiveness of surveillance has been slow to come to the attention of the public. After all, surveillance technologies have been designed and developed to blend in with everyday experience in such a way as to become almost invisible. Those who oppose the proliferation of surveillance have had to find dramatic ways to render opaque that veil of transparency. For example, the New York Surveillance Camera Players perform short dramatic works on the sidewalks in front of surveillance cameras for the benefit of those monitoring them as well as for the public. Another group, Applied Autonomy, has created an electronic map of the locations of all the surveillance cameras in Manhattan. Using a program called iSee, you can enter your current location in Manhattan and your desired destination. The program will return with a route designed to guide you to your destination encountering the least number of cameras. Computer software like Anonymizer erases your tracks online, thus allowing you to avoid cookies (i.e., commands hidden in a computer that automatically activate links to other sites) and other forms of Internet monitoring.

We raise the issue of surveillance technologies to begin this chapter because it illustrates the need for understanding technology in terms of the concepts of articulation and assemblage. If you were to approach this topic from the received view of culture and technology, you would be left with a wholly inadequate picture of what is going on and extremely poor tools for influencing or changing the role of those technologies. Often, surveillance is treated as a purely technological question: The problem is the technology and its effects. Typical questions include: What is the impact of using surveillance technologies? Should there be cameras or not? Does face-recognition software work? This mechanistic, often technological deterministic, view cannot account for the reasons for the development of surveillance technologies to begin with, or for their interpenetration in everyday life. When the origins of surveillance technologies are considered, they are typically done so in a cultural determinist, often expressive way: identifying the single cultural reason for their development. Some cite national security, especially in the wake of September 11. Some cite increased crime. Some point to the increasing isolation of individuals in contemporary culture, a situation that leads to suspicion. Others point to a growing culture of fear, especially fear of those who are different, sometimes referred to as "the other." Some draw attention to the new forms of commerce that require more sophisticated marketing techniques. It is as though, at best, the causal tools that feel familiar lead us to find an explanation, including praise or blame, in either the autonomous technology or an autonomous cultural cause.

We assert that the technology alone cannot explain the myriad ways in which surveillance matters in everyday life. Nor is there is any single reason that explains the rise in the number of cameras or surveillance technologies. Rather, there are multiple dimensions that need to be understood in order to get an adequate grasp of the place of surveillance technologies in contemporary culture. Articulation and assemblage provide tools to understand these dimensions, and open up useful strategies for action in relation to surveillance technologies, indeed in relation to any technologies, with far more sophistication and hope of being able to make a difference. Articulation draws attention to the contingent relations among practices, representations, and experiences that make up the world. Assemblage draws attention to the structured and affective nature and work of these articulations.

This approach is nonmechanistic and thus operates with the three nonmechanistic assumptions discussed in Chapter 9. The assumptions are: (1) that technologies are not autonomous, that they are integrally connected to the context within which they are developed and used; (2) that culture is made up of such connections; and (3) that technologies arise within these connections as part of them and as effective within them. However, this approach differs from expressive causality in significant ways. Primarily, while it does hold that culture is made up of connections, it does not insist that all these connections are reducible to an essence or attributable to a critical factor. Rather, culture is understood to consist of corresponding, noncorresponding, and even contradictory practices, representations, experiences, and affects. Note this last term: affects. We do not refer to effects, as in the outcome of a causal process, but to affects as a state: as disposition, tendency, emotion, and intensity.

Technology as Articulation

Perhaps the crucial thing to understand about articulation is the assertion that culture is made up of articulations (or connections) that are contingent. Contingency implies that these articulations or connections are not necessary, and it is possible that they could connect otherwise. In explaining how articulation works, Stuart Hall once used the image of a truck.[4] Imagine a semi with a cab and a trailer. The cab is articulated (connected) to the trailer. Together they constitute a connection, a relation, an articulation, and a unity: a truck. But this connection is not necessary. It is possible to disarticulate the cab and the trailer and rearticulate it by attaching a different cab or a different trailer. The newly configured truck is a new identity and a new unity, even though it too might still go by the name "truck." All identities or unities are like this: they are made up of articulations, but these articulations are neither necessary nor permanent. Identities are thus contingent; in other words, they are dependent on the articulation of particular elements that could change, thereby changing the composition of the identity. *Articulation can be understood as the contingent connection of different elements that, when connected in a particular way, form a specific unity.*

But what are these "elements" that get connected? The answer to this requires rethinking the term "element," which is misleading in that it suggests only "things," like cabs and trailers, or computers and video cameras. However, elements, understood as articulations, can be made of words, concepts, institutions, practices, and affects, as well as material things. Indeed, one can articulate an idea to an object to an affect, like connecting "progress" to automobiles to the affect "cool." In addition, every so-called element is itself an articulated identity, and therefore always part of a connection of still other "elements." A car is a unit, but it articulates many elements: parts, processes, a manufacturing industry, roads, advertising, an ideology of individualism, the pleasure of speed, and so on. The idea of progress seems to be a simple concept, but it too is made up of many other ideas, practices, and affects: a belief in evolution, the valuation of industrial technologies, the pleasure we take in gadgets, and so on. So rather than draw attention to the articulation of things, a cultural studies approach draws attention to the movement and the flows of relationships. Because language and popular philosophy have "taught" us to talk about and understand the world in terms of things, we do still tend to think and talk about things. But the challenge is to remember that even things are merely labels for momentarily frozen elements (misleadingly) isolated from the web of contingent relationships within which they are animated. Culture is better understood as the movement and flow of relationships within which things are created and animated, rather than as the accumulation of things.

We propose that you think about technologies in terms of articulations among the physical arrangements of matter, typically labeled technologies, and a range of contingently related practices, representations, experiences, and affects. Thus, surveillance technologies in the United States post-9/11 would be understood as being the particular contingent relationships among (at least) the following: the physical arrangements of matter (such as the thing we might call the video camera); the fear of terrorism; the propensity to think of space as something that needs to be controlled; a desire to care for and protect citizens; the belief that cultural profiling can predict and prevent terrorism and crime; the acceptance of a level of racism, classism, and sexism; a popular culture that idolizes new technology as "cool"; the titillation typically felt when snooping in a culture in which much is kept private; a strong commitment to the technological fix; a belief in the equation of new technologies with progress; the existence of a physical infrastructure and knowledge necessary to produce increasingly complex technology; a global intelligence community; a governmental leadership that emphasizes a particular political agenda; a legal practice that operates within a framework of rights and laws that define privacy within particular parameters, and so on.

These articulations are not fixed for all time; they do not remain permanently in place. They can and do change over time. But, here too, the speed and direction of change is contingent. Some articulations remain relatively tenacious; they are rather firmly forged and difficult to disarticulate. Hall called these "lines of tendential force," which draws attention to their tendency to remain articulated.[5] Others, however, might be more easily broken and thus subject to disarticulation and rearticulation. It all depends on the particulars of the nature of articulations at any particular historical moment. For example, on the one hand, legal efforts to protect the privacy of citizens, given their articulation to a political commitment to the rights of individuals expressed in the Bill of Rights, might be successful in reshaping the legal framework of what constitutes unjust invasion of privacy and effectively curtail certain forms of surveillance. On the other hand, the cultural commitment to the technological fix, to the equation of new technologies with progress, and the pleasure of the "cool," are not likely to go away any time soon. The tenacious force of this latter commitment will work against disarticulating the use of technologies in any form, most certainly against curtailing the "cool" use of surveillance technologies.

To think of technology as articulation insists, as should now be obvious, that technologies are not mere things. Rather, they consist of complex articulations that have been typically thought of as the context of a technology. But as you can see, one of the insights of articulation is that context, or culture, is not something "out there" out of which technology emerges or into which it is put. Rather, the particular articulations that constitute a technology *are* its context. There is no culture *and* technology; rather there is *technological culture*.

Technology as Assemblage

Technology as articulation draws attention to the practices, representations, experiences, and affects that constitute technology. Technology as assemblage adds to this understanding by drawing attention to *the ways that these practices, representations, experiences, and affects articulate to take a particular dynamic form*. The concept of assemblage is drawn from the work of philosophers Gilles Deleuze and Félix Guattari in their book *A Thousand Plateaus*.[6] Although their understanding of assemblage is more richly philosophical than the version we present here, it is still a powerful concept in this somewhat scaled-down version.

The concept of assemblage might best be understood by thinking about the term "constellation" used by Deleuze and Guattari when they talk about assemblage. A constellation of heavenly bodies like the Big Dipper, for example, takes a particular form: It selects, draws together, stakes out, and envelops a territory. It is made up of imaginative, contingent, articulations among myriad heterogeneous elements. The constellation includes some heavenly bodies and not others; these bodies only appear to be in proximity with one another given a particular act of imaginative gathering and the angle of our view across space. And, since both they and we are constantly moving, the relationship and angle changes. Further, the particular collection of (moving) bodies is articulated to a particular image: a dipper and not, say, a cap.

The constellation of our example could be said to *territorialize* the articulations of heavenly bodies, angles of relationships, space, atmospheric conditions, trajectories of movement, and a way of seeing. It is, in a sense, a contingent invention, both artificial and natural. However, once drawn into this form, the constellation exhibits some tenacity; it doesn't simple appear and disappear. The constellation that is called the Big Dipper has been called that for a very long time. Further, the constellation matters, in that it has real effects on our lives: It is effective in terms of practices. For example, the practice of astrology relies on the designation of constellations. It is effective in terms of representations. For example, we teach our children to read the sky in terms of constellations. It is effective in terms of affective experience; we feel at home in the hemisphere where constellations are familiar. As Deleuze and Guattari put it, assemblages are "constellation[s] of singularities and traits deducted from the flow—selected, organized, stratified—in such a way as to converge ... artificially and naturally."[7] For Deleuze and Guattari, an assemblage involves an intermingling of bodies, actions, and passions. In this sense, then, *an assemblage is a particular constellation of articulations that selects, draws together, stakes out and envelops a territory that exhibits some tenacity and effectivity*.

To this point we have talked in terms of the elements drawn together as practices, representations, experiences, and affects. But it might be helpful to expand a little on this list using terms from Deleuze and Guattari, who suggest that what is drawn together is both *forms of content* and *forms of expression*. Content includes what they call the "machinic assemblage of bodies, of actions and passions, an intermingling of bodies reacting to one another." These bodies can be, of course, both human and nonhuman, heavenly and mundane. Expression includes what they call the "collective assemblage of enunciation, of acts and statements, of incorporeal statements attributed to bodies."[8] Thus, whether the cultural theorist looks for connections among practices, representations, experiences, and affects, or between forms of content and forms of expression, both acknowledge the work of the material and the imagined, the lived and the represented.

A technological assemblage will obviously select, draw together, stake out, and envelop a territory that includes the bodies of machines and structures. But it also includes a range of other kinds of bodies: human bodies, governmental bodies, economic bodies, geographical bodies, bodies of knowledge, and so on. It also includes the kinds of articulations listed in the previous section: actions, passions, practices, commitments, feelings, beliefs, affects, and so on, such as those that we argued give shape to the identity of surveillance technology. In making the leap to technological assemblage, it is important to remember that a technological assemblage is not a simple accumulation of a bunch of articulations on top of one another, but a particular concrete constellation of articulations that assemble a territory that exhibits tenacity and effectivity. Thus, we may be able to characterize a surveillance assemblage post–9/11. To characterize that assemblage, we would have to do more than list its elements. We would be charged to "map" the territory with attention to the power of this constellation to assemble specific bodies, actions, passions, and representations in particular ways, to give a world shape, so to speak, in a concrete and imaginative way.

Let us demonstrate with a different example. Over the last few years, self-service checkout lanes have been introduced in supermarkets across the country. Rather than standing in line for a checker to scan and weigh items, bag them, and take payment, the customer can now stand in line to scan and weigh their own items, bag them, and pay the machine. This is said to be more convenient. Perhaps it's even said to be progress. Be that as it may, the received view would look at the situation as the machines merely appearing in the supermarkets and then having effects: unemployment for store workers, increased employment for equipment-repair folks, varied states of satisfaction and dissatisfaction of customers, increased or decreased profits for store owners and equipment manufacturers, and so on.

But to think the self-service checkout assemblage, we have to begin mapping the articulations. This means not only mapping the invention, design, and distribution of the machines, but a whole constellation of bodies, some of which are machines. For example, we need to consider the form and practices of the self-service machines and how they relate to, or resonate with, pay-at-the-pump gasoline, ATMs, vending machines, self-service machines in libraries, and the appearance of self-service machines in other types of stores.

Beyond the physical machines, there are other bodies and articulations in this constellation. Beyond the more obvious role of desire for profits and the delegation of labor to machines in the name of progress, we should consider the idea of self-service itself, and its articulations to ideas such as convenience and to do-it-yourself practices such as pumping your own gas, pouring your own drinks, and bussing your own table in a fast-food restaurant. Why do we do these things rather than have someone do them for us? We would also need to consider the articulation to the process of training customers and employees. People must be taught how to use the machines, but also *to* use the machines in the first place. Both these practices—to use and how to use—require training in new habits and practices. Customers have to be taught a whole new attitude toward purchasing, and a whole new model of how to purchase. The expectations of the consumer and their relationship with store personnel must be dramatically altered. The assumption that a customer is "waited on" must be disarticulated, and the customer must be convinced that this is a convenience, a good thing, a pleasurable activity, and so on.

So when we consider the self-service checkout-machine assemblage, we have to consider a whole array of machines, practices, habits, attitudes, ideas, and so on, which reach far beyond the effects of physical machines on cultural practices.

Assemblages do not remain static, however. Rather, they are characterized by a constant process of transformation: of what Deleuze and Guattari call *deterritorialization* and *reterritorialization*. *Deterritorialization* describes the process by which some articulations are disarticulated, disconnected, unhinged so to speak. *Reterritorialization* describes the process by which new articulations are forged, thus constituting a new assemblage or territory. The transformational process is virtually guaranteed by the myriad articulations that are subject to change. Sometimes rearticulations can contribute to reterritorializing an assemblage in significant ways. Sometimes the differences are effectively inconsequential. For example, it is clear that the surveillance assemblage post–9/11 is not the same as the surveillance assemblage pre–9/11. While many of its elements, taken in isolation, "look" the same, the overall assemblage has changed. A video camera post–9/11 may "look" just like a video camera pre–9/11, but it is not the same from the perspective of the technological assemblage. In contrast, some of the technical features of video cameras post– and pre–9/11 may "look" dramatically different; but these differences may be relatively insignificant from the perspective of assemblage. When people get excited about the appearance of a new technology and begin to prophecy its effects, they may be missing the possibility that in terms of the effectivity of the assemblage overall, nothing really significant at all may be changing.

The argument we are making clearly connects with the insights about agency we raised in the previous chapter. As we argued there, it is not technologies or people that have and exercise agency. Rather agency—the ability to bend space, to make something happen—is possible or not possible depending on the particular assemblage. That assemblage may or may not assemble the world in such a way that agency is *attributed* to one thing or another. It just so happens that the assemblage within which we find ourselves—the technological culture of North Americans in the twenty-first century—assembles technological practices, technological representations, and experience in such a way that we tend to think and feel that technology is a causal agent: the bearer of progress, the deliverer of convenience, the guarantor of the good life, and so on.

Rearticulating Technological Culture

So what is there to do if you want to change the culture? What practical strategies follow from understanding technology as assemblage? The first lesson is to be certain that your analysis has been of the technological assemblage, and not of the technology as thing. If you don't like what you see, don't blame technology or the culture; understand the assemblage that maps technological culture.

Then if you want to imagine or contribute to change, look more closely at the particular articulations that account for the particular constellation of the assemblage. Where are there lines of tendential force, those articulations that you may not be able to disarticulate? There you may not be able to accomplish much. As Hall has written, "if you are going to try to break, contest or interrupt some of these tendential historical connections, you have to know when you are moving against the grain of historical formations."[9] But also consider where there might be lines, connections, relationships, and articulations that could be altered. Where might the topic of a college class matter? Where might a legal case make a difference? Where might saying no to a particular technology be significant? To answer questions such as these requires careful analysis of an assemblage, and how in that assemblage the particular bodies we call technology fit. Thus, to build on the example of surveillance we developed earlier, it would probably be far more successful to work toward curtailing the growing pervasiveness of the surveillance assemblage by appealing to a commitment to the right to privacy guaranteed in the Bill of Rights, rather than trying to convince people that their blind love affair with new technology is serving to erode their privacy. The commitment to the Bill of Rights hits home affectively in the mainstream heart of the United States, even if real understanding of those rights is limited. And that affective response can articulate to the work of law, building on legal precedent to craft ways to protect privacy. Such a strategy is likely to be far more effective than trying to convince people to give up their unquestioning acceptance of technology. In their classes, Jennifer and Greg have been trying to convince people to question their assumptions about and relations to technology for years. You tell us: How successful have we been? And how successful do you think such an appeal could be more broadly? And what might the mechanisms be for getting that message out there?

Of course, this is not to suggest that education doesn't matter or that contesting lines of tendential force is futile. Indeed, this is not to argue that you should seek only the paths of least resistance, only the weakest articulations, but that you become aware of the tenacity of whatever you are trying to change. So, for example, surveillance likely will not be significantly affected in North America until one addresses the decline of community and the isolation of individuals that leads to lack of trust, suspicion, and fear of others. That is a monumental task, but one worth addressing in our opinion. Given the complexity of any technological assemblage, one can never be certain about what processes of rearticulation might make a significant difference. Sometimes the world throws curves, and the work of complex articulations that we haven't noticed before comes screaming on the scene to remap the territory in significant ways. There were those few who predicted a terrorist attack on the magnitude of 9/11, but most people thought that possibility was out of the question. Once that attack happened, however, the surveillance assemblage took a turn few of us would have predicted. Similarly, processes of rearticulation can work for good ends: hence the commitment on the part of Jennifer and Greg to write this book. This book is testament to the belief that rearticulating people's understandings of the relationship between culture and technology away from the idea that they are autonomous entities and toward the idea of technological culture or the technological assemblage can make a difference, even if that difference is down the road a ways. Sometimes the rearticulation of small matters will connect with larger ones, and the world changes. As Deleuze once put it, "Our ability to resist control, or our submission to it, has to be assessed at the level of our every move."[10]

Conclusion: Why Articulation and Assemblage?

Technology as articulation and assemblage offers a whole new way of posing the "problem" of culture and technology. No longer is it possible to think in terms of either technological determinism or cultural determinism, or for that matter, some hybrid of the two positions. By understanding assemblage, flow, relations, connections, and articulations as what matter, the "things" themselves, the physical arrangements of matter, drift into helpful perspective. They are not unimportant; they are just no longer all-important. They do not act alone or independently. Assemblages—those imaginary yet concrete constellations—matter. To understand their structure, their work, their power, and their reach, is the *task* of the cultural theorist. To contribute to changing them in constructive directions is the *goal* of the cultural theorist.

CHAPTER TWELVE

Space

PICTURE YOURSELF IN A MODERN AIRPORT, hurrying to catch a plane—it doesn't really matter which modern airport, or which plane. The airport terminal is crowded with passengers jostling toward their gates or chatting excitedly as they head toward the baggage claim. The human traffic is a flow, much like a river, with eddies and rapids. The flow in the central concourse works its way around lines of passengers waiting to embark, streams of passengers disembarking, customers lining up for coffee or fast food at kiosks and counters. There are pockets of passengers standing almost in the middle of the flow and staring upward at a row of monitors that announce arrivals and departures.

Consider this latter group. If the monitors had not been there, this group would not have stopped at that position, however briefly. If the monitors had been placed elsewhere, the traffic pattern would shift, altering how one could move down the concourse. The monitors seem to emit a force field that affects the passersby. The monitors contribute to the shaping of movement through this space and the activities that occur there. The information provided (flight numbers, times, gate numbers, and status information) contributes to the pattern; but the material means, medium, the thingness of the technology, also contributes enormously. Large wall-sized boards found in some train stations provide similar information, but given a whole range of other spatial and architectural considerations, contribute differently to how those spaces are shaped. Television, the technology in question in our airport example, crops up in a variety of settings, both public and private, and interacts with the spaces and flows in ways that make it matter.[1]

Technology Is Spatial

Throughout this book we have grounded our discussions of technological culture in examples drawn from everyday life. By focusing on everyday life, our discussions consider technologies as they occur and are used within particular spaces.

For example, it is more revealing to focus on cell phones in everyday life than to speculate about cell phones in general. By conjuring images of that phone on a belt, in a purse, at home, at work, in the car, in a classroom, or on the street, it becomes clear that technologies are an integral part of the particular spaces we live in, work in, walk in, and drive through.

Everyday life is predominantly spatial, and the spaces within which we exist are always cultural. If we focus on technologies as spatial, our assumptions about technology change. By looking at communication technologies as spatial, as we do below, we demonstrate how technologies can articulate space and time differently. Further, a spatial approach to technology highlights significant ways that our bodies are implicated in technological culture, thus reinforcing the importance of framing a theory of technological culture within the context of everyday life.

Concepts of Space

Before we get to any of these issues, let us clarify what we mean by space. Space is usually used to mean a physical space. However, we draw on the concept of *social space*, which asserts that space is the production of social relations over time. That's the condensed version. Here's a slightly expanded version drawn from Greg's book *Exploring Technology and Social Space*:

> What I mean by social space can perhaps best be approached negatively: it is not merely a constructed space like a room in a house or a lobby in a hotel or a city street. Likewise, it is not merely the meanings generated by any single human moving through that space (i.e., that the room seems comfortable, that it reminds one of corporations, that the greens in the wallpaper seem soothing, that it is a workplace or a home, private or public, that he or she is in a hurry, at ease, looking for the bathroom, etc.). Social space is the space created through the interaction of multiple humans over time. There is never a single social space, but always multiple social spaces. Social spaces are always open and permeable, yet they do have limits. It is important to remember at this point that the social is not unique to humans. Baboons, insects, and other creatures are social and could be said to move in social spaces.[2]

We have seen in Chapter 10 (on agency) that if we want to understand any particular concrete local situation (an airport concourse, a classroom, and so forth), we need to consider how both human and nonhuman actors contribute to the specific shaping of that space. Social space is shaped on many levels. French sociologist and philosopher Henri Lefebvre offers three interrelated concepts of social space. The social space we live in and move through is the combination or articulation of all three. These three concepts are *spatial practice*, *representations of space*, and *representational space*. We can think of these as space as it is *perceived*, *conceived*, and *lived*.[3]

Spatial practice consists of the structures and activities that produce and shape space by articulating it in certain ways. Architecture is perhaps the most obvious form of spatial practice, but so too are practices such as the recurrent behaviors of the people in particular places, including such behaviors as hanging out, walking along a sidewalk, sitting at a computer terminal, and reading the monitors in airports or train stations.

Representations of space—space as it is conceived—refers to the concepts we use to think about space. This, as Lefebvre explains, is the dominant space of our understanding,[4] and includes the ways that scientists, engineers, planners, architects, and others understand and represent space as something to be lived. Maps, blueprints, architectural plans, rulers, and light-years are all representations of space that have a relationship to a variety of spatial practices.

Representational space is the direct, lived bodily experience of space, which includes how we move in, through, and experience space. It is our awareness of space as we variously accept, appropriate, and change space as a lived experience in the intersection of spatial practices and representations of space. It is what space "feels" like.

The particular arrangement of articulations among these three concepts in any particular space is what we designate a *cultural space*: a space unique to a particular way of life, articulation, or assemblage. For example, to explore the space of the airport concourse and all those television monitors, we would have to consider it along these three dimensions. How do the televisions, the architecture, and the flows of traffic contribute to shaping the space? What are the attendant concepts of this space? Airports are conceived as very technological spaces and tend toward the sleek and modern, so ideas of progress, modernity, and speed might shape our approach to and understanding of this space. What are our experiences of this space? How are they shaped by our readings of the semiotics of the space, the colors, and the signs? How are these spaces shaped by all those film and television shows that have been set in airports and airplanes? What does it feel like? After September 11, 2001, this means that issues of terrorism and security are part of this space, and so we may experience the airport in a dramatically different fashion than before those events. This way of thinking about space is a way of getting at the role of technology in everyday life.

Let us give you another example. On the campus where Greg teaches, the primary classroom building is shaped in a large square "C" around a courtyard and fountain pool next to some grassy areas. This area is a space; it was constructed to have a certain affect, but by itself it is not a cultural space. As the sun gets lower in the early evening, students gather in this space in preparation for evening classes. At this time in the evening, it is one of Greg's favorite spots on campus, because students and professors from many walks of life come together here. They gather in groups, chat about life and class, and grab food and drink from the small snack bar on the ground floor. It feels like community happening. The space that this courtyard has become is produced by the many paths crossing it, activities that take place in it, and meanings that are given to it by those who pass through.

However, over the last two years, as more people have adopted the use of cell phones, the shape of the space of the courtyard has shifted. People still come and hang out, eat, drink, and talk before class, but many of them are not talking with each other anymore, but to their cell phones, and to people in other places. The sense of this space has thus expanded to include others who are not present. In so doing, the space changes in character, because people are not interacting in a face-to-face manner with each other as much. Here is this courtyard, it now feels less like community than a gathering of individuals.

The example above is one way that cell phones have contributed to the shaping of space. By connecting an individual in a particular place with another (or others) in different places, the experience of space includes this extra dimension. Howard Rheingold has written about what he terms "smart mobs," collections of individuals who keep in constant or regular contact through the text-messaging feature of their cell phones.[5] These collections of groups, be they friends, protestors, or business colleagues, exhibit a form of organization that is quite fascinating. What we wish to point out here, about this practice of messaging a small circle of friends, is that the importance of these messages lies less in their content (most are along the lines of "what's up") and more in the fact that they maintain the relationships between individuals as they move differently through different parts of the city. They are about maintaining relationships in a particular spatial form.

Technology and Everyday Life

The spatial presence and awareness of technology pervades our everyday lives. Everyday technology is not just those technologies we can think of as being at hand, although this is certainly a rich field of study that includes the cars, pens, cell phones, and microwaves we use on a daily basis. Everyday technology also includes those technologies that operate more on the periphery of our experience, but nonetheless participate in the spatial character of everyday life. These include airplanes overhead, the potential of weapons of mass destruction designed to invade or defend space, the penetration of pollution from the technologies of heavy industry, and the changing space of farms in relation to the changing technologies of food production and distribution.

A growing number of studies have begun to look at technology on the everyday level, and many of them focus on the reconfiguration of the domestic sphere. Ruth Schwartz Cowan's study of housework, which we have mentioned before, is germinal here. Others have studied television, VCRs, and computers in this way.[6]

For example, Lynn Spigel's history of the early days of television, *Make Room for TV: Television and the Family Ideal in Postwar America*, focuses on how television was involved in the transformation of middle-class family space, especially the space of the home. She writes:

> Home magazines primarily discussed family life in language organized around spatial imagery of proximity, distance, isolation, and integration. In fact, the spatial organization of the home was presented as a set of scientific laws through which family relationships could be calculated and controlled. Topics ranging from childrearing to sexuality were discussed in spatial terms, and solutions to

domestic problems were overwhelmingly spatial: if you are nervous, make yourself a quiet sitting corner far away from the central living area of the home. If your children are cranky, let them play in the yard. If your husband is bored at the office, turn your garage into a workshop where he'll recall the joys of his boyhood. It was primarily within the context of this spatial problem that television was discussed.[7]

The question posed by these magazines was: Where in the house should the TV set go? And where in a room? They noted that when a television was placed in a living room, it should be the focal point; the rest of the furniture was to be oriented around watching it, especially chairs and couches.[8] The TV set also displaced other household furniture and features: It became the new hearth around which the family gathered, sometimes literally placed in front of the hearth, and quickly replaced pianos as the centerpiece of entertainment.

Spigel writes that by the early 1950s, house floor plans published in magazines included the television as part of the home's structural layout. "Indeed, magazines included television as a staple home fixture before most Americans could even receive a television signal, much less consider purchasing the expensive item."[9] Television was said to bring the family closer into a romanticized image of the family circle. By watching TV together, husbands, wives, and children could have an activity to share. Television was also said to help keep delinquency down by keeping children off the streets and at home. Spigel points out that the emergence of the "family room" as an architectural feature of the post–World War II suburban home indicates the importance placed on bringing the family together.

In India, the domestication of television takes place in a different cultural assemblage. Neena Behl studied television use in India where Sundays became "TV day" and Sunday night focused on television viewing.[10] This new ritual changed how food was prepared so as to maximize time spent watching television. It also brought together men, women, young, and old in a common activity, where previously these groups typically spent leisure time separately. However, when viewers were from different families, the arrangement of people in the space in front of the set still differentiated them by social rank, gender, and caste.

The significance of television is not limited to its role in changing domestic spaces, but in public spaces as well. Anna McCarthy discusses the placement of television in public spaces such as stores and restaurants. She looks at such things as the placement of televisions, who can see them, and what sort of programming is on them. She writes:

> Such issues may seem dictated by transparent functionality, but they are simultaneously, like all architecture, forms of social communication. The position of the TV set, in short, helps to position people—not necessarily the empirical persons who work, wait, and relax within eye- and earshot of a particular screen but certainly the spectator positions these persons are encouraged to occupy within the social organization of the space and within larger networks of power, as well.[11]

It matters in a restaurant, bar, or store if only customers, and not employees, can see the television (and vice versa). It matters if the content is advertisements, or if the content is related to the business or not. In each of the variations, different spatial practices, representations of space, and representational spaces give unique shape to the technological culture.

These examples illustrate that all three concepts of space are crucial to understanding technological culture. Practices of technologies are spatial. Representations of technologies are spatial. Also, technologies are integral to the experience of spatial relations.

Additionally, these examples bring to the surface the *embodied* dimension of technology and everyday life. Embodiment is addressed more fully in the coming chapters, but let us make two points here. First, we need to acknowledge the physical work performed by technologies. Sometimes a technology shifts the space around itself in significant ways. Sometimes a technology takes up space and needs to be accommodated. For example, other technologies might need to be shifted out of the way: a TV replaces a fireplace; a bread machine displaces a blender; a computer displaces a typewriter; and an e-mail displaces a letter. Further, technologies sometimes contribute to reshaping human bodies and human movement in space. The shape of a keyboard influences how arms and hands move, a sidewalk steers people along particular trajectories, high-heeled shoes render certain kinds of movement more difficult than others, and so on. We see the consequences of this shaping in many ways, including technologically derived complaints of muscle strain, carpal tunnel syndrome, eyestrain, back pain, and so forth. The recognition of the power of the embodied shape taken by a technology gives rise to the field of ergonomics, a field of research concerned primarily with movement and spatial relations concerning human–technology interactions.

Second, we must acknowledge the everyday *affective* dimension to the embodied nature of technological culture. The intense emotional responses we have with, through, and against technologies are all embodied aspects of experience. The frustration at a malfunctioning appliance or slowly functioning procedure indicates how closely our bodies are bound up with our technologies. Although we often think of emotions as purely mental constructs in the brain, *affect* is an embodied physical response *of the body*. The mental realization that one is angry is different than the embodied experience of anger. The power and significance of the embodied emotion, the affective response, has been a neglected aspect of most discussions of technology. The feeling of the sublime discussed in Chapter 1 is relevant here, as is the mini-sublime: the "cool" and the "neat," the contemporary articulations of the sublime. The experience of a new technology might not fill you with a grand sense of the sublime, as it clearly once did for many people, but it might fill you with a sense of the "cool" or the "neat."

However, affect plays a role beyond the obvious ways in which humans are entangled emotionally with their technologies. The extent of that role is difficult to see, because technology is largely understood to be an offshoot of science and therefore "rational" and "practical." Emotions and affect are viewed as the opposite of rational. As Rosalind Picard pointed out in her study of affective computing, even such common personality tests such as the Myers-Briggs promote this assumption that thinking and feeling are opposites.[12] In contrast, Picard argues that recent research into the brain suggests that all perception and conception (thinking) run first and foremost through the limbic system, which is the seat of emotions. Picard's argument is that computers need to be able to deal with the affective aspects of human life. There are at least two reasons for this. The first is to aid in human/computer interactions. If a computer can recognize your emotion, it might respond more appropriately: "She keeps clicking her mouse; she must be getting frustrated with my error message; I should respond differently." Also, if the computer can present information with an affective component, humans might be better able to evaluate and process it: A computer might deliver a happy message, an urgent message, or a desperate message. Second, if computers are "thinking machines," and if we have discovered that thinking involves emotions, that is to say, embodied reactions to situations, then our computers will not think very well until we have determined how to embed them in the physical (and cultural) world, and until they process information based on a variety of situational, embodied inputs. Indeed, studies of patients who exhibit physically decreased emotional response, perhaps due to a brain injury, have determined that these patients find themselves almost incapable of making rational decisions.[13] And yet we wish to trust machines that are solely "rational"?

Modes of Communication

In communication studies, there is a unique, if somewhat problematic, body of literature that looks at the articulation of space, technology, and everyday life. This body of literature is sometimes called the *modes of communication* argument. Although modes of communication verges perilously on technological determinism, it introduces insights that merit a closer look. Generally, the argument holds that culture has moved through several "modes" of communication—several whole new ways of life that have been given definition by particular technological ways of communicating. Three modes central to the argument are: orality, literacy (script and print), and electronic (or digital—it is always most difficult to name the mode you find yourself in). While this argument has been developed in much greater detail elsewhere, we present just enough here to highlight the integral connection between technological culture and space. Specifically, we look at oral, literate, and electronic media and arguments made regarding their integral connection to space.

Orality

We begin, somewhat historically, with orality, the medium of the spoken word or spoken language. Language is, after all, a technology: it is a creative externalization that then requires reperception and engagement in a dynamic process of scaffolding.[14] In a culture that is completely unfamiliar with writing—a condition called *primary orality*—sound and the spoken word take on a particular shape and importance with implications for concepts of time and space.[15] As Walter Ong, Jack Goody, and Ian Watt have explained, sound goes out of existence just as soon as it comes into existence. In other words, sound does not hang about. It has no permanence. Once a sound is uttered, it expires and cannot be called back.

Most communication in oral culture is face-to-face, since the human voice can only carry so far, which makes most social relations personal and direct. To communicate in this culture, one must pay attention, because there is no going back unless someone repeats what was just said. Repetition is, however, a separate utterance, not an exact copy of the first. The practice and art of repetition, in fact, become very important in spoken language in order to aid memory. Memory is paramount in oral culture because once something is forgotten by all members of the culture, it is really gone. There is no looking it up later. Because it is easier to remember narrative and rhythm than discrete facts, important information, such as when to plant crops, is typically "stored" in stories, rhyme, or song. Further, if one is in a group and someone is talking or singing, everyone in that group shares that experience; they hear at the same time. The sound is unavoidable.

Given this reliance on the spoken (or sung) word in order to remember, there is very little room for independent thought in a primary-oral culture. Consider this: If one has an original idea, how does one remember it? Consequently, primary-oral culture is characterized by a group-oriented mindset and cohesion. Ritual storytelling is essential and there is an emphasis on tradition. Consequently, the reach of face-to-face interaction and the maintenance of group cohesion define the space of oral culture. Those who remember (such as elders or designated keepers of stories) are revered and powerful, but only insofar as they contribute to group cohesion.

Harold Innis, a Canadian economist writing in the mid-1900s, described primary oral cultures has having a *bias toward time*, rather than a *bias toward space*.[16] A bias toward time means that the culture maintains cohesion by exerting control over time, that is, by managing memory, keeping and telling stories, in such a way that time is collapsed into a perpetual present. To claim "that is how we have always done things" or "if you do *x*, *y* will happen, because it has happened that way before," are ways of perpetuating order and cohesion by obliterating the difference between what we would now understand as past and present. Time biased cultures operate within a limited spatial context, dependent on the sound of one's voice. The small tribal unit characterizes the optimal space of a primary oral culture; it has little concern to exert control beyond that space of one's voice. The shift in balance from time bias to space bias is dramatic, with the transition from oral culture to literate culture.

Literacy

Once a culture adopts writing and reading, things change dramatically. The crucial development is the separation of the message from the sender, which permits the message to move independently in space. This brings into being the distinction between forms of expression and forms of content. Messages can now travel to those not present at their utterance, which rearticulates the role of memory (of the faulty memory of the messenger, for instance) and processes of distortion (of a message as it passes from person to person, for instance). Communication with the imprimatur of authenticity can now reach across a greater space, thus expanding the influence of a culture. Leaders can control empires by sending messages across long distances. Collapsing space, another way of thinking about controlling space, becomes a critical concern of these cultures; hence they are considered to have a space bias.

Simultaneously, the ability to control time diminishes. With writing and reading, a (relatively) permanent record of communication minimizes the need to memorize. Instead, the maintenance of collective and personal memory becomes a matter of storage in physical form, and archives become important repositories and sources of power. The ability to consult records and multiple versions of stories breaks down the coherence of a collapsed time. There are now pasts, presents, and possible futures, and those who give shape to the records, and have access to them, hold power. The ability to read and write, as well as to create, maintain, and access libraries of information, becomes a potent resource. Knowledge becomes centralized in the hands of the powerful few with the foresight and might to compile, control, and interpret it. Once detached from its form of content, the form of expression is open to new possibilities for interpretation.

Group cohesion diminishes, as does reverence for elders, when memory is no longer important, when memorization and performance are no longer group activities, and the form of expression can be detached from the content. Individual creativity, personal isolation, and independence become the standards of literate cultures. The ability to store information outside the body frees space in one's mind for creative thought. Unlike attending to sound, the act of reading is fundamentally a solitary experience. One reads to oneself, even if one is in a group. Even while reading the same passage to themselves, readers give the text their own meaning and inflection. In literate societies, what is called the *cultural time lag* exacerbates this situation. This lag is a literal one, marking the lag in times between when something is written, disseminated, and read. Further, since people do not all read the same things or may read the same things at different times, there is less in common and thus less social cohesion. Thus, when compared to primary oral cultures, literacy tends to isolate people rather than draw them together.

Because not all written communication is the same, the significance of the balance between time and space biases varies among literate media. Innis pointed out the profound differences that various writing media can make on the structure of a society. For example, the ancient Egyptians carved their writings on stone, a rather enduring medium. These carvings were meant to last for ages, passing on the traditions, knowledge, and religion of the rulers for "eternity." Therefore, this medium has a relative bias toward time. Since the writing process was slow, and the written materials not easily moved, power tended to be centralized and concentrated in the hands of the rulers who controlled access to the sites where the writings were stored.

The invention of papyrus changed matters. Papyrus was a lightweight paper-like medium made from reeds woven together, crushed, dried, and pounded smooth. Papyrus was easier to write on and could be readily carried from one part of the realm to another, enhancing the bias toward space. These factors supported significant changes: The pharaohs could communicate accurately and swiftly with distant parts of their realm, thus facilitating the extension of their power over greater distances and making empire possible for the first time.

With papyrus available in greater quantities, and an increased demand for writing, the scribal class expanded, beginning a slow movement toward universal literacy. However, papyri quickly crumble, which means that their writings were not meant for the ages. Something closer to universal literacy became possible with the mass production of paper and eventually the cheap presses of the nineteenth century.

Print Culture

Printing technologies constitute a significant later stage in the literate mode of communication. The invention of the printing press with interchangeable metal type (first invented independently in Korea and then more famously in Gutenberg's shop in Germany) contributed to a variety of cultural transformations. Written language became standardized, the volume of printed materials increased, and the price of books decreased. These transformations allowed for printing in vernacular (local) languages. Prior to this, books were mainly written in Latin, the language of the Catholic Church. The Church had virtually monopolized book production and the education of literate people throughout the Middle Ages. By printing in vernacular languages, the printers challenged the Church's authority and broke the monopoly on printed knowledge. Benedict Anderson has even argued that print languages (as he calls them) were a crucial component in the rise of the modern nation-state, what he calls "imagined communities."[17] Print languages also contributed to a greater sense of history in modern society. Since handwritten language changes so much in style and form over generations, older manuscripts can become quickly illegible. In contrast, a standardized print language set in metal type allows us to read books printed more than 500 years ago.

Let us sum up briefly before moving on. When we only communicated orally, our space was small, localized, and centered on the group. Written language allowed us to individualize our experiences of space and expand it, because we could read about distant events and communicate with and control distant peoples. Our sense of space, then, was no longer limited to the spaces we experienced immediately, but included far-off and imaginary spaces known only through writing.

Electronic Communication

The current mode of communication has been called the electronic mode, but more recently it is sometimes referred to as the computer era, the digital mode, or the Information Age. Regardless of what name is finally adopted, the communication medium that marks the new mode is the electronic telegraph. During the 1830s there were various experimental forms of electric telegraphy, although it was not until Samuel Morse developed his code (Morse Code) that the technology was rendered cost effective.[18]

Communicating via electrical signal was virtually instantaneous, allowing one to be in immediate contact with someone else hundreds of miles away. Though now an accepted part of our everyday routines, this required a fundamental shift in spatial practice, representations of space, and representational space. Prior to electric communication, messages only traveled as fast as the transportation technologies that carried them (ship, horse, wagon, and so on). In fact, communication and transportation were virtually synonymous. The primary change that the telegraph affected, James Carey and John C. Quirk have argued, is the separation of communication from transportation.[19] Instantaneous communication thus altered the structure of space. In so doing it also altered the economy by restructuring national stock and commodity markets. As Carey explains, telegraphy made it possible to level markets in space, because prices could be easily compared. Thus markets shift to speculating on futures, because the effects of change over time could not be controlled.[20]

The telephone compounded changes in space by separating the sound of one's voice from the speaker. Because hearing someone's voice conveys a sense of presence or nearness, the telephone brought those far away into a form of spatial nearness, what is now called telepresence. Notice how we sometimes say—with just a touch of wonder—to someone on the telephone who is otherwise far away, "you sound like you are just next door." This collapsing of space continued with the advent of radio and television. Marshall McLuhan, a student of Innis, predicted in the 1960s that we were coming to live in the space of a *global village*. He argued that instantaneous global communication, via satellite and submarine cables, puts us almost literally in each others' backyards, thus forcing us to be as concerned with one another as neighbors are. However, McLuhan ignored the ways that sudden proximity can also cause enmity among neighbors.[21]

The coming of electronic communication contributed to yet another change in our practices and experience of space. It shifted communication away from the eye (in terms of reading) and back to the ear. We have entered a stage of what Walter Ong has called *secondary orality*, in which orality is shaped and sustained by technologies that depend on literacy, including printing and electronic technologies. In a secondary oral culture, we attend aurally as well as visually to communications aimed at groups; mass broadcasting is the archetype technology. Ong argues that this leads to a strong group sense that is more powerful in its size and in its ability to extend across space than is the group sense of a primary oral culture. Furthermore, in secondary oral cultures, individuals have the choice of independence or group belonging.[22] One may be group minded, but can be self-consciously so. This is not the case with primary oral cultures, where group belonging is a necessity of life. The possible dangers of secondary orality are exemplified in the rise of fascism in the 1930s, driven in part by the effective use of radio as a tool of propaganda. The essence of fascism is to draw a large population together around a core group identity.

The latest development in electronic communication and space is the establishment of the Internet and its rapid and intrusive growth in many forms. These new technologies have not only contributed to a shift in space by speeding up the cycle of global communication, but they have also ostensibly created a new space, *cyberspace*, that exists solely on networks of cables, routers, and computers. The way that cyberspace will contribute to the changing technological culture is emerging around us. Many have described how it expands the space of communication, enhances forms of contact, and allows multiple, global spaces to be created. But since computer use is most often an individual activity, we are also becoming physically isolated and distanced from those who are otherwise close to us. This tension between distributed interpersonal connectivity and physical isolation is evident in the example of smart mobs that was discussed earlier. Interestingly, however, smart-mob technology has been used to enhance the ability of groups to physically congregate for social or political reasons, so the emerging situation is rich in complexities in its rearticulating of space.

Conclusion: Why Space?

Thinking of technology in terms of space provides at least three insights, which we need to carry forward. First is to emphasize one of the primary themes of this book: We cannot consider technologies in isolation; they exist and function in, and as a part of, spaces that are cultural. To ignore these dimensions is to seriously misunderstand how technologies work in everyday life.

The second insight is an appreciation of our own experiences. Technologies are not disconnected from the space of our everyday lives, and any analysis of technology should be able to start from where we actually live. Technological assemblages are not just about industrial technologies, factories, offices, battlefields, and so on. They are also about the practices, concepts, and experiences of everyday life. An appreciation of our own experiences also emphasizes the insights from previous chapters, specifically the contingency of any technological assemblage: Technologies—their formations, uses, and effects—are contextual. Every encounter with a technology, every delegation and prescription, is a particular articulation and as such is never guaranteed. We should not discount our own experiences simply because they go against the general rule of how a technology is "supposed" to work.

The third insight to be gained from thinking about technology in terms of space, and a consequence of the second point, is the realization that other articulations of technology are possible for us and for others in other spaces. If we recognize the contingency of our practices, representations, and experiences, we might begin to question our own assumptions about how we feel things are supposed to be, and call into question the times when we simply accept an assemblage as somehow "natural." We can also recognize that others in other spaces may have practices, representations, and experiences of technology at odds with our own, and that these too are based on complex articulations that we may need to understand. The degree to which these other articulations may be commensurate or incommensurate, bridgeable or unbridgeable, needs to be studied. It is empowering, we would argue, to realize that the world could be different for us and for others.

CHAPTER THIRTEEN

Identity Matters

DOES IT MATTER WHO OR WHAT A PERSON is when it comes to issues concerning technology? Our answer is an unqualified yes. As we have argued, a technology is not a neutral tool. Rather, technologies are developed and used in circumstances, and for ends, that do not treat everyone equally. Identity—who or what someone is—matters in a range of ways:

- Technologies are developed to meet the needs of some and not others.
- Technologies are not distributed evenly. There are technology haves and have-nots.
- Technologies make assumptions about who will be using them and how they will be used.

In general, identity affects how a person is placed in culture: how important they are, how they are treated, and what possibilities are open to them. To illustrate, imagine the difference in placement of someone identified as a judge as opposed to someone identified as a criminal, or the placement of someone identified as rich as opposed to poor. In an ideal, utopian world a person's identity would not be met with prejudice or barriers to social achievement, but that does not describe the world as we experience it. In fact, what technologies are made available to a person depends in part on their identity (gender, race, class, ethnicity, citizenship, ability, and so on); how well integrated a person is into the prevailing technological culture likewise depends on their identity; and technologies are used to assign (sometimes challenge) identity. Identity is not just an issue *after the fact*—a matter of distributing technologies to some and not others—but an integral part of the technological assemblage. It is an issue that permeates the whole of technological culture, that is, the ways that people are situated in relation to technology throughout the practices of everyday life.

In this chapter, our argument is organized around three main propositions: First, that technology is unequally delegated along identity lines. Second, that technology is unequally prescriptive along identity lines. Third, that what we call "technologies of identity," which assign or reassign identity, not only create and alter identity, but reinforce or challenge particular notions of identity, including the questions of what it means to exist or to be human. In the following chapter we specifically take up the question of challenging identities.

But first a note: When we discuss identity in these chapters, the term encompasses a wide range of identities. We mean it both, in the most obvious sense, as our place within cultural categorizations (gender, race, class, ethnicity, citizenship, ability, and so on) but also, in the less obvious sense, as our existence (an ontological question of being or not being).

Technology Is Unequally Delegated

As we argued in Chapter 10, following the work of Bruno Latour on agency and Actor-Network Theory, we do delegate tasks to technologies. Technologies are made to stand in place for human actions, but not every task is delegated. Choices are made at some point by real people and real organizations about which projects deserve support and which technologies should be developed, distributed, and used. For example, in science and engineering, some research projects are supported with research funding and others are not. Further, even if a technology is developed, real people and real organizations make choices about how and to whom technologies will be distributed and used. Technologies will be designed to work in ways that suit those uses and users.

The connections between identity and technology are complex and pervasive, and they begin in a sense with the circumstances of where and how technologies originate, and who or what is responsible for their appearance. While it is beyond the scope of this book to tackle the historical particularities of this question as it applies to North American culture, our discussion of definitions of technology in Chapter 8 offers some clues as to how to address the question. We submit that technological innovation comes from many sources: the irrepressible human urge to create, the personal needs and desires of everyday life, the potential for fame and profit, the concerns of communities for social welfare, the idea of "the good life," and the institutional demands to survive and grow, to name a few. It is safe to say, given our particular North American context, much of the innovation of what we generally recognize as new technology (new technological *products*) occurs within the context of institutional research and development. These institutions may be government (military, university) or corporate (for example, AT&T's fabled Bell Labs or Microsoft's R&D).

In this context we can consider what criteria are used to "greenlight" (as the film studios say) a technology. Profitability is often the main criterion, but there are others, including efficiency, convenience, speed, and competitive potential. These criteria are supported by value systems that privilege the needs and desires of some populations over others. These criteria will greenlight technologies that enhance the lives of some and diminish the lives of others. We offer an example of the discriminatory nature of these criteria by focusing on speed.

In the 1970s Ivan Illich argued in his book *Energy and Equity* that speed discriminates.[1] Those able to go faster than others are assumed to be more important. Automobiles are built for and represent a speed class; and those who own one are considered more important than those who do not. (Who is typically considered more prominent and imposing, the person who owns an automobile or the person who does not?) Further, those who own faster automobiles enjoy status not given to those who own slower ones, even if they don't drive them fast. (Who is typically considered more prominent and imposing, the person who drives a Porsche or the person who drives a Yugo?) It is almost as though the potential for independent speed is what is valued. Those who rely on public transportation, which tends to be slow and less convenient, are severely devalued in American society (though there are isolated exceptions).

While enhancing the lives of some, the discriminatory nature of speed devalues the lives of many. Speed, Illich argued creates scarcity, hierarchy, and exploitation. He explains it thus:

> More energy fed into the transportation system means that more people move faster over a greater range in the course of every day. Everybody's daily radius expands at the expense of being able to drop in on an acquaintance or walk through the park on the way to work. Extremes of privilege are created at the cost of universal enslavement. An elite packs unlimited distance into a lifetime of pampered travel, while the majority spends a bigger slice of their existence on unwanted trips [to and from work]. The few mount their magic carpets to travel between distant points that their ephemeral presence renders both scarce and seductive, while the many are compelled to trip farther and faster and to spend more time preparing for and recovering from their trips.[2]

If speed supports hierarchy and exploitation, how fast is too fast? At what point does speed begin to discriminate between those who are privileged and those whose lives are devalued? According to Illich's calculations, we get into trouble when the average speed of a society exceeds a mere fifteen miles per hour. As Illich succinctly puts it, "Tell me how fast you go and I'll tell you who you are."[3]

Now, one would hope that when decisions were made about delegating tasks to technologies that the goal of providing the greatest good for the greatest number would guide those decisions. This is not always the case. Technologies created in a cultural context are shaped by the values and conflicts of that culture in many ways. They will most often support (at least some version of) the status quo. At the very least, no organization will willingly create and promote a technology that eliminates the need for that organization; this is the principle of self-preservation. As we saw in the example of speed, technological delegation supports the elites, including those who own the means of production.[4] Benefits such as public safety, which depart from the primary criteria for delegation, are often incidental, an afterthought, or brought about through other means. An example of delegation brought about by other means, in this case through engaging in technological politics (the topic of Chapter 15), is the consumer advocacy movement founded by Ralph Nader in the 1960s, which forced the automobile industry to reluctantly add (just) the most rudimentary of safety features to cars such as seat belts.[5]

The process is more complicated than we have set out here, and there are always exceptions to the idea that technologies support the status quo. It should be clear that our approach does not reduce culture to the expression of a single causal set of practices and beliefs. Rather, decisions are made by many people with different agendas, values, and criteria, all of which might variously contribute to shaping technology and the technological assemblage. For example, in a marvelous and creative book titled *Aramis, or the Love of Technology*, Bruno Latour presents the case of technological innovation as a detective story.[6] He asks, how did Aramis—a plan for a fully automated train system—come about, get shaped, funded, refunded, revised, advocated, and finally killed? It is a fascinating and complex story with dozens of characters and plenty of plot twists.

Complexity does not make the process of innovation democratic, however. What is most common among all the processes of technological innovation is precisely this fact: They are nonparticipatory. Not everyone, with their various agendas, values, and criteria, participates equally in making the decisions that matter. Technological decisions are made by some and then impinge on others. In response to the prevailing undemocratic process of innovation, Richard Sclove argues for the value of strong democracy in making decisions about technology.[7] We discuss his ideas more thoroughly in Chapter 15, but touch on a relevant point here. The principle of strong democracy, which he borrows from Benjamin Barber, is simply this: Citizens should have input into the decisions that affect their lives. Sclove argues that because technology most definitely permeates one's life, the individual should have a say in how that technology is created, designed, and implemented. He writes:

> This model of democracy, even in schematic form, is sufficient for deriving a simple but compelling theory of democracy and technology: If citizens ought to be empowered to participate in determining their society's basic structure, and technologies are an important species of social structure, it follows that technological design and practice should be democratized. From strong democracy's complementary substantive and procedural standards, we can see that this involves two components: Substantively, technologies must become compatible with our fundamental interest in strong democracy itself. Procedurally, we require expanded opportunities for people from all walks of life to participate in shaping their technological order.[8]

At this point we begin to hear grumbling from our readers that we have missed a crucial and democratic element of the story: the thumbs-up/thumbs-down process of deciding the fate of new technologies: consumer choice. The argument goes like this: We live in a market economy where the consumer is sovereign. Consumers do not passively accept every product they are offered. The poor reception of the Edsel, New Coke, and a host of other unsuccessful technologies offer proof of that. Consumers have a need, and a technology is created to meet that need. If there is no need for the technology, it will not sell. Admittedly, from time to time manufacturers come up with a technology to meet a need that we never knew we had, but this is all to our benefit.

This is an old debate and one that, if pursued, would take us far off track. But for our purposes here, let us make three points about the consumerist argument of technological development:

- This account of technology ignores the role of technologies (as devices) in the shape and effectiveness of a technological assemblage. It especially ignores the ways that individual consumer decisions are shaped and influenced by the assemblage. On the one hand, new technologies participate in the articulation of new needs. On the other hand, our needs are shaped by forces of which technology is already a part. For example, I may not want to own and use a computer, but I may feel that I need to, given a situation in which everyone else I deal with uses computers to communicate via email.
- This account assumes that consumers have an unlimited range of technologies to choose from, which is simply incorrect. Consumers can only choose from the options provided. This is akin to shopping in a grocery story; you can only choose to purchase what has been put on the shelves. Some technologies are never offered to consumers because they do not meet criteria for technological innovation, such as profitability, convenience, speed, or efficiency; but we might have chosen these if they were offered.
- This account has a very narrow and uniform view of consumer needs. Different groups in society may have radically different needs. Some may need a new DVD player, while others may need housing or clothes. "But," the consumerist argument responds, "wouldn't separate market niches be created for each group? We know that we are not a homogenous society and that mass marketing and mass production are passé." Our answer: "Not necessarily." If one group has money and another does not, whose needs will be catered to?

This last point leads us to a final example of the unequal delegation of technology that addresses the conflicting values and practices that drive technological development. The example comes from the world of medicine, a realm in which we would dearly hope that the fundamental criteria at work would prioritize the health and welfare of humanity in general. Unfortunately, it is not working out that way.

We begin this example with a widely accepted need: a cure for sleeping sickness.[9] Sleeping sickness is a disease found in Africa, caused by a bite from a tsetse fly. At one time the disease was almost wiped out, but recent decades of war and economic catastrophe have brought it back. The disease threatens millions of people; in some portions of Africa it is more prevalent than AIDS. Like AIDS, sleeping sickness is 100 percent fatal, but unlike AIDS, sleeping sickness is 100 percent curable. The current cure is an incredibly painful series of injections of a caustic drug (a combination of arsenic and an ingredient that goes into making automobile antifreeze). The drug itself kills 1–5 percent of the patients. But there is an alternative.

The alternative is a drug called DFMO, which was originally created by a pharmaceutical company to cure cancer. It didn't cure cancer, but it did cure sleeping sickness without side effects. Here is the problem and the conflict in value systems: DFMO costs around $600 per patient because it is expensive to produce. This would be no problem for patients with money or insurance, but the people most affected by this disease cannot afford it. Affording the drug is a moot point, however, because the pharmaceutical company stopped manufacturing it since it would not be profitable to do so. The drug could save millions of lives, and alleviate much pain and suffering, but no one will manufacture it. The pharmaceutical company, Aventis, did give the formula to the World Health Organization, but the WHO could not find anyone willing to make it. What this example shows is that profit, rather than social good, is a dominant criterion of medical innovation today. Typically tasks will be delegated to new medicines (and research money spent to pursue these new medicines) only if they are expected to be profitable ones.

Our conclusion here is reemphasized in the second chapter of the DFMO story, when manufacture of the drug was resumed, but not as a cure for sleeping sickness. It turns out that DFMO helps in hair removal and is now being marketed in a new hair-removal cream, Vaniqa, for well-to-do women in the West. It was still not made widely available to save lives in Africa.

The third chapter in this story opens in May 2001, when, under pressure from Doctors Without Borders (Médecins San Frontières: MSF) and the WHO, the pharmaceutical company Aventis committed to again make available this drug and others to combat sleeping sickness. Bayer had also made a similar commitment. On the one hand, this again illustrates the importance of engaging in technological politics. On the other hand, it is now 2004 and we have had little success tracking down information about just how available these drugs have actually become. How available do you suppose they might be?

As this case illustrates, the process of delegating some tasks to technology and not others integrally involves identity. The process of delegation is hardly democratic. To fully appreciate the consequences of delegation, and the pressing need to democratize it, it is important to understand how technologies are unequally prescriptive. This, in turn, will enrich an awareness of how identity is implicated in matters of technology.

Technology Is Unequally Prescriptive

Just as the delegation of tasks to technology benefits some parts of the population over others, technologies are unequally prescriptive. Those who delegate to a technology are not the only ones impinged on by it; once a technology is in place, it acts on all those who encounter it. Though this sounds egalitarian, remember that technologies are designed with a set of assumptions about its users: what they weigh, how tall they are, their abilities, their intelligence, and other demographic factors. The technology, in turn, assumes that all users match this profile, in spite of the fact that they will not all do so. Technologies make two kinds of assumptions: what we will call *design assumptions* and *system assumptions*. Both figure significantly in matters of identity.

First, design assumptions are the basic assumptions of an individual technology about the people using it. These assumptions usually go unnoticed unless you are someone for whom the technology was not designed. If you are left-handed, you know exactly what we mean: ladles pour from the wrong side, doors close from the wrong side, handle grips feel awkward, and writing on most school desks is a challenge. In another example, very tall, very large, or very short people can easily spot the limitations of automobile design. Most automobiles are built to accommodate "average"-sized people, discriminating somewhat against women, whose average height tends to be shorter than men. Seatbelts don't fit, the mirrors are placed wrong, and the steering wheel is at the wrong angle, to list a few problems. Modern automobiles allow users to adjust the seat, steering column, and mirrors, but only within a limited range of possibility.

Most computers are constructed with the assumption that you have the use of both hands to type and use the mouse, and that you can see the screen. A lot of software, and much of what is available on the Internet, assumes that you read and understand English. In addition to this, by their very design, computers shape how we use them. Beth E. Kolko writes that, "technology interfaces carry the power to prescribe representative norms and patterns, constructing a self-replicating and exclusionary category of the 'ideal' user."[10] But computer design makes assumptions about more than just the physical and mental abilities of this ideal user. In fact, computers assume an awful lot. The computer assumes that you have a consistent and reliable power source, the money to pay for that power, a place to put the computer, a phone line, reliable phone service, the money to pay for that service, the money to purchase software, the time off from work and other activities (making food, raising children) to learn how to use the computer, and so on. It also assumes that you have access to someone who can fix it when it breaks and update it when it's obsolete. Computers assume, in short, much more than is accurate about the majority of the world's population. But this discussion has moved us from a consideration of design assumptions to the second kind of assumptions that technologies make: of *system assumptions*.

Because technologies are not isolated tools but parts of systems of technologies, we can detect the work of system assumptions: the assumptions made by the system within which the technology functions. We can begin by looking at support systems (fuel costs, replacement costs, repair costs, supply costs) and how some technologies are connected to others. Fast-food systems provide an interesting and extended example of the consequences of systems assumptions. Think of the McDonald's restaurant chain as a technology, or rather as a series of technologies connected together in a technological assemblage, the goal of which is to provide inexpensive hot food to a maximum number of people in a minimum time.[11] This system works quite well for most of the population. As sociologist Susan Leigh Star puts it:

> McDonald's appears to be an ordinary, universal, ubiquitous restaurant chain. *Unless* you are: vegetarian, on a salt-free diet, keep kosher, eat organic foods, have diverticulosis (where the sesame seeds on the buns may be dangerous for your digestion), housebound, too poor to eat out at all–or allergic to onions.[12]

If McDonald's recognizes a significant market demand, it will alter its system to cater to these particularities: vegetarian burgers in some places, mutton burgers in others. However, no matter how many niche markets one identifies, there will always be people outside the system who are inconvenienced or even harmed by the system. Star is allergic to onions, even if cooked, and finds that getting any food establishment (not just McDonald's) to omit the onions (to not even put them "on the side") an endless task. Greg is lactose intolerant and needs to police his food for milk products; *you* try ordering a pizza without cheese and see the looks you get. However, lactose intolerance in recent years has been more widely recognized as a common disorder and has become a niche more readily catered to than has onion allergies. Many airlines have limited or eliminated their distribution of peanuts as a classic in-flight snack because of acknowledged concern over the fact that people with peanut allergies can have severe life-threatening reactions from being exposed to peanut particulate matter in the air. This is again the work of technological politics.

Star raises the point about being allergic to onions, not simply to draw attention to the potential cruelty and inconvenience of large-scale systems, or even to remind us that there will always be people outside the system, ignored by design practices, though these are important points. She emphasizes instead that we are all affected by such technological systems, that we must begin with "the *fact* of McDonald's no matter where you fall on the scale of participation, since you live in a landscape with its presence, in a city altered by it, or out in the country, where you, at least, drive by it and see the red and gold against the green of the trees, hear the radio advertising it, or have children who can hum its jingle."[13] Technological systems thus impose and *impinge* on people who do not even consciously participate in those systems. To emphasize more broadly how technological systems prescribe (or impinge) unequally, and less personally, we turn to examples of how technologies impinge along a number of highly significant cultural dimensions: gender, race, class, and ability.

Gender

In Chapter 2 we introduced Ruth Schwarz Cowan's classic study, *More Work for Mother*, which traces the history of what have been called labor-saving technologies in the homes: electric dishwashers, clothes washers and dryers, refrigerators, vacuum cleaners, and appliances in general. The "purpose" of these technologies was to accomplish strenuous tasks with less effort and time, and make life easier. Although all of these technologies were developed by men working outside the home, they affected women in the home. Since the tasks addressed by these technologies have traditionally been women's tasks, these technologies impinged on women much more than on men. What Cowan discovered is that these technologies actually *increased* the amount of time women spent laboring in the home. They did not save labor, time, or effort for these women. Instead, the men's inventions placed more demands on women and their work. How did this increase happen? Many tasks once outsourced to others, such as laundry and ironing, could now be done—and therefore had to be done—at home. Tasks that once included children and other family members—in such family efforts as "wash day"—could now be done by one person, and inevitably that one person was the mother. Furthermore, these technologies contributed to creating a higher standard of cleanliness than had been previously appreciated and expected. Because we could now conveniently and easily launder clothes on a daily basis, the technology contributed to the notion that we *must* launder them after every wearing to be considered clean. Carpet cleaning had once been a communal annual or semiannual activity for the family. Carpets were rolled up, hauled outside, and beaten. With the introduction of the vacuum into the household, the carpet could suddenly be vacuumed and cleaned more frequently, a fact that contributed to the belief that it *must* be vacuumed and cleaned more frequently. Sweeping the floor was no longer good enough. Now carpet cleaning with a vacuum is a solitary activity that can be performed far more often, even weekly or daily.

The technologies certainly did not achieve these effects "on their own." The *gendering* of specific tasks (delegating tasks to one gender more than another) is definitely reinforced by advertisements and social expectations. A technology is always developed in, or rises out of, particular circumstances, and its use is always introduced into a gendered environment. Ann Gray, for example, found that when VCRs were introduced into homes in Britain in the 1980s, their use matched established gendered patterns for the use of household technologies.[14] Specifically, men tend to use technologies at home for specific limited tasks, such as fixing a leak, making bookshelves, or changing an automobile's oil; whereas women tend to use technologies for ongoing day-to-day chores, such as housecleaning. In addition, "high-tech" devices tend to be male territory.[15] Although leisure technologies in general tend to be gender-neutral, Gray found that it was the men who usually learned how to use the VCR first and remained the household experts on more advanced functions such as timed recording. Though the women in her study learned how to record, playback, and rewind a tape, they usually turned to their male partners if the machine needed to be programmed to record at a later date and time. Control issues are especially evident in the observation that if more than one person is watching television, the remote control is almost always in the hand of a male adult, or, if not, then a male child.[16] Further, the VCR became an element in a long-shifting negotiation of leisure space and time within the domestic environment, an environment in which it is much easier for men to establish "time out" for relaxation than it is for women, who are constantly concerned with domestic chores.

A similar set of issues arose when the personal computer became a fixture of many homes in the 1980s and 1990s. Marsha Cassidy describes how the personal computer was marketed predominantly to men in the 1980s, but with slowing sales and a desire to make PC's a common domestic appliance, marketers began targeting women in the 1990s.[17] Following Cowan's work, Cassidy points out that the PC was advertised as a labor-saving device, in particular saving the labor of women in the home. It did so by emphasizing the PC's role in allowing women to work from home (telecommuting), manage the household (keeping track of shopping lists and family schedules), and enhance her children's education. Like other domestic technologies before it, the PC also makes more work for mother by increasing her responsibilities (and potentially moving the working mother back out of the office and into the home, concentrating paid and unpaid work in the home). Cassidy's research also touches on the spatial dimensions of technology addressed in the previous chapter, in that she raises the question of where to put the PC in the home. Because there are few spaces in a typical home that are exclusively a woman's (versus a man's "den" or the children's own rooms) the location of the PC and the gendered responsibilities for its use raise key questions about gender, labor, and technology in the domestic environment.

Race and Class

To begin to address the ways in which technologies impinge unequally on the basis of race and class, we turn to an example discussed by Langdon Winner: the case of Robert Moses, the architect who designed some of the major public works of New York from the 1920s through the 1970s.[18] Moses was responsible for parkways, bridges, and other large constructions that we often take for granted. Indeed, we often do not consider such structures as technologies, though of course they are. One might assume that public works impinge on all users equally. After all, how can a road discriminate? Can't we all drive on them equally? Ah, there is a lesson here in how we can be so easily deceived by the appearance of things.

We focus on one of Moses's public works: the bridges over the parkways on Long Island. These overpasses are amazingly low, in some places leaving only nine feet of clearance overhead. This does not hinder anyone driving a standard automobile, but the bridges effectively hinder the passage of taller vehicles, like trucks or busses. Therefore, the bridges discriminate against those who drive trucks or ride busses. Those who ride busses are less likely to own their own cars and more likely to come from the lower class. Consequently, the lower classes have a more difficult time getting to Long Island. Minorities are also more likely to make use of public transportation, so their access is restricted as well.

Are these bridges an unfortunate mistake or a thoughtless error? According to Moses's biographer, Robert A. Caro, the bridges were deliberately designed to hinder poor people and blacks, not only from using the parkways, but also from accessing Jones Beach, a park Moses designed. In this case, the task delegated to the technology was in part that of racial and class discrimination. The bridges continue to impinge this particular lesson back on all who drive (or who cannot drive) down the parkways of Long Island.

A key issue throughout the 1990s, as the Internet was booming, was the increase in the digital divide. The digital divide refers to the unequal distribution of computer technology, as well as access to computers and the Internet. This unequal distribution (based on race and class) would lead, it was argued, to a society even further divided between the haves and have-nots. There are any number of reasons for such a divide: The poor cannot afford computers, computer training, or Internet access fees; poor schools cannot provide students exposure to, much less training on, computers, to name a few. The solution to the divide is usually predicated on a call for universal access: Wire every school and public building (especially libraries) to make sure that everyone can get online. Even the homeless can get online through public terminals.

There are limitations to the universal-access argument that become clear when we talk about it in terms of delegation and prescription. In terms of delegation, we have to consider who delegated what to computers and the Internet? There are recent studies of race that discuss how much of the Internet was created by white males, and how it also reflects the assumptions of white males about what that technology should do.[19] The Internet technology then presupposes (this is prescription) that the users share certain experiences and culture as well as socioeconomic standing. Access debates typically assume that those who gain access will use the medium in similar ways for similar reasons. But what is needed (in terms of representation of the experience of multiple cultures and races online) is minority participation in creating the Internet and Internet culture rather than prescribing that minorities use what others have made.[20]

Ability

Technologies often impinge unequally on those with different abilities. Most technologies assume that the users are able bodied. Computers assume that you can see and type, curbs assume that you have legs rather than wheels to move you, and books assume that you can read. For those who are differentially abled, technological culture can challenge their sense of their very right to be! In some cases adaptive technologies are developed to make the technology useful for the differentially abled. In terms of adapting computer technology to facilitate surfing the Web, devices can enlarge images and text on the screen, read the screen text out loud to the user, or display the text in Braille on a special reader. However, despite federally suggested guidelines, most Web sites still assume that a user can see them.

In discussions of variable ability, questions are usually raised about norms and the extent to which the differentially abled should have to adapt to what is considered normal ability (seeing, hearing, walking), rather than having the "normal" world adapt to their differences. An example of such a debate is taking place in the deaf community over implant technologies, which could enhance or restore hearing. While there are many advantages to being able to function more readily in a culture that presupposes one can hear relatively well, there are those who argue that such devices undermine the identity of a unique culture that has been created among the deaf, who may live and communicate quite well without being able to hear. Hearing devices assume that a person who cannot hear is somehow less of a person than is someone who can. Rather than demanding that the deaf hear (through technological means), can we not accept them as they are? Whereas both sides in the debate tend to agree that pursuing such an implant is an individual choice, the question remains whether parents should make such decisions for their children.[21] As this example illustrates, we ought, at the very least, be cautious about making assumptions about the "appropriate" place of the differentially abled and assure them a place at the table in making decisions regarding technological delegation.

Technologies of Identity

Thus far, this chapter has emphasized how identity gets caught up in technological relations of delegation and prescription. We wish to extend this idea to a less obvious but even more potent level by considering two ways in which the relationship between identity and technology takes a more personal, corporeal turn. First, we consider the ways that identities are shaped in and by scientific schemes of categorization, such as the seemingly self-evident categories of gender or race, and by scientific scrutiny. Second, we consider the ways that technologies are used to mark and alter the body in order to enhance one's identity markers (for example, to appear more feminine or masculine) or to change them (for example, to lighten skin color or change gender).

Technologies of Categorization

One of the legacies of the eighteenth-century European Enlightenment was the idea that modern science was rational, ordered, and morally superior. It was believed that the world could be understood through rational means, by detached objective observation, and by the labeling and categorization of all things.[22] Everything was said to have a distinct identity and to be related to other things in distinct ways. The grand schemes of scientific nomenclature derive from this era. For example, Linnaeus attempted to categorize all living things in terms of kingdom, phylum, genus, species, and so on. As rational and logical categories, these divisions were thought to be absolute: You were a plant or an animal, not both. There were no gray areas. Among the most obvious categories were male and female, though not all species made this distinction. In terms of humans, the male/female division seems self-evident. But this is not always the case for humans. For example, a number of children are born each year bearing some physical characteristics of both sexes, and occasionally quite distinctive physical attributes such as the "wrong" genitalia. These children are referred to as being "intersexed."

This is where technology steps back into the picture. The scientific schemes for knowing and labeling the population become technologies of standardization and normalization, techniques for identifying the normal and the deviant through medical inspection. A child is declared normal or deviant, and those declared deviant have to be normalized. In the case of intersexed children, this can be as simple as purposefully ignoring the difference if it is slight, utilizing techniques to socialize the child "properly," or treatments as complex and radical as hormone therapy and corrective surgery. The chances of a child being intersexed in some way is one in two thousand (or about 65,000 per year).[23] Until a few years ago, it was standard medical procedure to immediately perform corrective surgery on the infant—without even informing the parents.

The presence of the "deviant" would suggest there is actually a continuum of body types from traditional male to the traditional female (and not just a continuum of body types—there is a much wider variety of chromosomal pairings beyond the XX and XY that most of us were taught in school). Cultural and medical technologies of normalization (and we mean those of categorization much more prevalently than those of surgery) work against that continuum and on the general population to identify, characterize, and reinforce discrete categories of physical characteristics and behavior. So when we identify ourselves as female or male, we do so as products of technologies of cultural conditioning and medical technique.

Like gender, race has been the object of intense scientific speculation and research. Despite the accepted scientific categories (Caucasoid, Negroid, and so on) there is no scientific, biological basis for racial differentiation. There are no physical traits that fall absolutely in only one category, and there is no DNA marker by which to differentiate the population. Racial categories and the characteristics attributed to different races are purely cultural.[24] However, this does not mean that schemes of racial categorization don't have real effects on real people. The technologies of racial classification have tremendous impacts on citizenship, immigration, and quality of life within different countries.[25] From access to jobs, education, and housing, to freedom of movement and rights within the legal system, racial classification systems have significant effects.

Technologies of the Body

In addition to technologies that are used by a society to impose identities on the population, technologies can be used to alter identities to either conform to or rebel against cultural norms. These technologies of the body range from makeup to surgery. Makeup is used to alter one's appearance to fit within cultural norms of attractiveness and to exaggerate or emphasize gendered characteristics of appearance, such as the eyes or lips. But makeup is also used to alter racial characteristics. For example, skin-lightening cream is used to change the color of one's skin so that it better meets the cultural ideal of fair skin and "white" identity. Other cosmetic technologies that work to alter identity include surgical technologies such as liposuction, collagen implants, breast augmentation and reduction, face-lifts, nose jobs, and penis enhancement. Women are the predominant users of these procedures, but men also use them. These surgeries can reinforce cultural standards of attractiveness.[26] Cosmetic surgeries also alter racial characteristics. For example, such procedures are relatively common in Southeast Asia, where Asian women have cosmetic eyelid surgery to rid themselves of their epicanthic eyelid to take on the rounder eye shape of Western (Caucasian) standards of beauty. However, this example is more complicated, since women may undergo the operation in order to minimize the racist reactions that their epicanthic eyelids elicit (as a marker of racial difference) rather than explicitly to look white.[27] As genetic science and technology become more sophisticated, the technology will be used to alter these identity characteristics on a genetic level by selecting out or altering the human genome.

Conclusion: Why Identity?

As this chapter has demonstrated, identity is deeply implicated in technological culture. Who and what we are is integrally articulated to the differential ways that tasks are delegated to technology, and to the ways that those technologies prescribe identity roles back on us. We are not some autonomous freestanding identity who develops and uses tools. Those so-called tools are, as Langdon Winner once put it, "forms of life," and we, as particular identities, interact with, give shape to, and are shaped in those forms of life.[28] However, just because people are not autonomous free-wheeling tool users, does not imply that they are completely subjugated to technology. Variably, people do resist the identities prescribed for them by technological culture, and they often use technologies to do so. In the next chapter, we turn to this practice of challenging identities.

CHAPTER FOURTEEN

Challenging Identity

IN DISCUSSING TECHNOLOGIES OF THE BODY in the previous chapter, we argued that identity is both a constitutive factor of technological assemblages, and a product of them. Plastic surgery and cosmetics technologies, for example, articulate to identity and are frequently used to bestow identity. In some cases, however, technologies of the body entail resistance; that is, they participate in practices of resisting identities otherwise bestowed by culture. For example, Anne Balsamo discusses how female bodybuilding challenges dominant notions of femininity by producing and promoting the ideal of heavily muscled women. Balsamo argues that such transgressions, though challenging, are often pathologized in the dominant culture, because they stray so far from the dominant stories of gender identity.[1]

Technologies can also challenge the notion of stable identity itself. Sex-change operations, for example, resist the notion that one's gender cannot change. In addition, because sex-change operations foreground a distinction between physical gender and psychological gender, the very notion of a stable gender is called into question.

This chapter examines the cultural work of challenging identities by taking up four particularly potent contemporary sites (articulations of a range of forces: practices, discourses, values, beliefs, and affects) where technologies participate in challenging the notion of stable identity: (1) online identity, (2) cyborg identity, (3) artificial intelligence, and (4) artificial life.

Online Identity

The seeming ephemerality of online identity has been a topic of much discussion since the 1990s. In her influential book, *Life on the Screen*, social psychologist Sherry Turkle interviewed students who spent a great deal of their time online.[2] She found that text-based interactive environments such as MUDs (Multi-User Dimensions), MOOs (MUD, Object Oriented), and even chat rooms allowed the students the opportunity to "be" someone else, occasionally several other people, because such environments are created solely by textual description. Online a person can describe their appearance, feelings, actions, and environment however they choose. They can be tall, handsome, well built, beautiful, funny, smart, and self-assured; they don't even have to be human. Beyond the initial description, they simply have to interact with others online according to their purported personality (confidently, quickly, intelligently, belligerently, with humor, and so forth). In engaging in these interactive role-playing scenarios over time, people often develop entirely different lives and identities for themselves. Some individuals run multiple characters on a single site, or different characters in different environments. Turkle quotes "Doug," a midwestern college junior:

> I split my mind. I'm getting better at it. I can see myself as being two or three or more. And I just turn on one part of my mind and then another when I go from window to window [on my computer screen]. I'm in some kind of argument in one window and trying to come on to a girl in a MUD in another, and another window might be running a spreadsheet program or some other technical thing for school.... And then I'll get a real-time message [that flashes on the screen as soon as it is sent from another system user], and I guess that's RL [real life]. It's just one more window.... RL is just one more window, and it's not usually my best one.[3]

Online experiences such as these have brought into the question the assumption that each of us has only a single core identity. Now, we didn't need computers to point out to us that we have many sides and aspects to our identity and personality. We only need to observe our own and others' behaviors in different situations to witness dynamically different personalities coming to the fore. For example, you might be focused and serious in class, but fun and flirty in a bar. But the Internet allows you to completely rework appearance in an online environment by controlling nonverbal communication. You are free to describe how you look, your expressions, posture, gestures, reactions, and so on. You can "try on" other appearances and personalities that would be impossible (or embarrassing) to carry off in real life. So this refiguring of identity goes far beyond dressing and acting differently for a particular occasion or event, where you would have far less control over the nonverbal aspects of who you are.

Recent gaming practices push the articulation between online and offline identities when the two are purposely designed to intersect. One night each week in the building where Jennifer works, a group of gamers with well-established online identities gathers to enact (act on, continue, and develop) those identities in real space in their real, corporeal bodies. The group is secretive about its activities, but the challenges to identity—when, for example, a corporeal man may be a woman both online and in the space of enactment—are fascinating. It is a strange thing to overhear or to stumble onto these games late at night, as sometimes happens. (One fellow faculty member described feeling "trapped in her office" during a lengthy and heated exchange outside her door in the early-morning hours.) But it is stranger still to encounter these "same" people walking the "same" halls in daylight during working hours. The seriousness of the challenge to identity is hardly captured by the term "game." Who *are* these people anyway?

The mode of communication facilitated by the Internet raises to new heights old questions about the cohesion of identity. Are we single selves or multiple beings? Can we change who we are? Online identities often seem independent and autonomous—just as cyberspace is sometimes seen as an independent and autonomous space—but chosen identities taking place in "real" space are not the same as "real life." However, from the perspective of assemblage, we have to consider the myriad articulations between online happenings and offline events and how identity is challenged in and by this reconfigured space. As Internet use becomes a daily activity for people, it is no longer a radically distinct, compartmentalized activity; and the play, experimentation, and activity that flows through these modes are surely sites of significant change in the shape of identity in technological culture.

Beyond challenging the notion of the single identity, online technologies also raise questions about many of the categories of identity that we tend to assume are fairly stable: those of gender and race in particular. Recently there has been considerable debate over whether or not a person really can leave these identities behind when they go online, or if they will inevitably reveal their identities in some form.[4]

It seems a commonplace—if not entirely correct—that we can ascertain the identity of someone just by looking at them, or perhaps just by hearing them speak on the telephone. We don't mean that we will discover exactly who someone is, but people believe that it is usually possible to figure out some major identity categories rather quickly: gender, race, age, and sometimes even class (plus there are markers of ethnicity and sexuality that could be read as well). As technologies come between us in our interactions, and if we bring those online identities purposely into the "real world," this ascertainment of identity becomes more difficult.

Carolyn Marvin, in her history of the early days of electric technologies, describes how the telephone presented a major social problem when it was introduced into the middle- and upper-class home. Since you couldn't see whom you were talking to, there was no way of knowing in advance whether it was someone that you were "supposed" to talk to or not. For example, it played social havoc if a woman talked to (or was found talking to) a man who was not well known to her and her family, to someone of the lower classes, or to someone of a different race. Such matters of social distance and the distinction between public and private affected the whole family, and beyond. Marvin writes,

> [T]he picture that emerges from less self-conscious accounts in the professional literature is one of the bourgeois family under attack. New forms of communication put communities like the family under stress by making contacts between its members and outsiders difficult to supervise. They permitted the circulation of intimate secrets and fostered irregular association with little chance of community intervention. This meant that essential markers of social distance were in danger, and that critical class distinctions could become unenforceable unless new markers of privacy and publicity could be established.[5]

With online technologies that present no analog dimension of communication—that is, one does not see them or hear their voice, but only reads their words or views the animated image that the other has created—much more than just the markers of social distance are challenged. All markers of identity are said to be erased online. When one is just looking at the words on a screen (be they in a chat room, interactive game, or Web page), one cannot automatically discern the race, class, gender, sexual orientation, or ethnicity of the writer. Indeed there are any number of ways of masking one's identity online and asserting anonymity. The appeal of this erasure of identity markers was captured and promoted in this advertisement for the long-distance telephone company MCI: "There is no race. There is no gender. There is no age. There are no infirmities. There are only minds. Utopia? No, Internet."

The notion that the Internet erases identity markers and is thus only mind-meeting-mind has bolstered the infamous practice of online gender bending. Gender bending occurs when individuals in online interactive domains describe and name themselves as being of the opposite sex. Some online environments even encourage participants to invent new genders, helping to deconstruct the cultural insistence on two and only two genders. Most often gender bending seems to involve the practice of men posing as women, perhaps because it was not until 2001 that the number of women online equaled that of men.[6] However, as Dale Spender found, women have learned the advantage of a gender-neutral or masculine screen name. In interactive environments, those with feminine names or who self-describe as female tend to get harassed or solicited for "cybersex." In contrast, when women present themselves online with gender-neutral or masculine names, they find their ideas taken more seriously and their positions debated on their merits. In other words, the experience of many (though by no means all) was that the Internet was only a level playing field of mind-meeting-mind when all involved were assumed to be male.[7]

The example of gender bending shows that even though the online realm has the potential to level the playing field, such is not necessarily the case in actual practice. Gender stereotypes and gendered behavior follow individuals online, even if given a new twist. That said, there are positive lessons to learn from the practice of gender bending; most notably, it can reveal to individuals firsthand the differential privilege accorded to gendered individuals in spite of claims of equal treatment.

Whereas gender has become an arena of play and creativity online, a much less talked about category in this context is that of race. Like gender, racial characteristics can be masked online and revealed or not through self-description or naming practices. But unlike gender, race is more absent than bent online. Some argue that one needs some sort of gender identity (even a new one) in an online environment simply for pronoun agreement (he, she, it, and other inventions). Because race, in contrast, does not impact language in the same way, there is less need to recreate race online; we don't formally switch pronouns depending on the race of the speaker or subject.

Given the violence and tension of centuries of racism, the Internet suggests itself as a true utopia, where people are equal—judged, as Martin Luther King Jr. said, by the "content of their character" rather than by the color of their skin.[9] However, online environments tend to go further than allowing for the erasure of race; they insist on it. People who reveal their race online are vulnerable to harsh verbal attacks (known as "flames") that declare them racist or accuse them of stirring up trouble.

There are two questions that need to be raised in probing the significance of such incidents. First, if one's race or ethnicity is part of one's identity and culture, why shouldn't one discuss and declare such things online? Why should participants deny a central part of their lives while online? Which brings us to a second point: Why such hostility? What is really going on with respect to race in this raceless space called cyberspace?

The treatment of race online demonstrates a subtle power of privilege that dominant groups have: the power of not having to be named or marked. Whenever one marks a difference, a hierarchy is assumed. So when one says, "There are two categories, A and B," culturally one implies, "one is preferred, the other is not; one is powerful, the other is not; one is normal, the other is not." For example, etymologically the word "woman" derives from the Old French for *wifmann*, *wif* meaning female and *mann* meaning human being.[10] Note than *mann* does not mean "a male human being," just "human being." In other words, the cultural norm is "man" and anything that is not man (for example, woman, child, animal, technology) is marked in relation to man. Man is said to be "the measure of all things,"[11] but remember that when you measure something you are judging the object measured, not the yardstick, even though the yardstick ought to be judged as well. One of the privileges of being dominant in society is to be the yardstick by which others are measured, and since one (incorrectly) assumes the neutrality of the yardstick, it doesn't need to be named or have attention drawn to it in order to proclaim its role and presence. If someone is "normal," there is no need to say so. But if someone is different, we mark this fact. For example, note how media typically name the race of a subject only if it is other than Caucasian ("an African American attorney"). Consequently, one of the privileges of being dominant in society is to not have to be marked.

This process of not being marked indicates a "cultural default." In our culture, lack of alternative information suggests that an individual conforms to the cultural dominant for that group. For example, "doctor" defaults to male. Sometimes people struggle not to be victim to the cultural defaults; often they will vehemently deny having any such preconceptions. However, there is much to be learned from the riddles that frequently circulate that are meant to trip people up when they rely on the cultural default. There is that moment of surprise, perhaps brief and fleeting, when Dr. Smith turns out to be a woman.[12]

It is an unfortunate fact of our culture, and one that needs to be contested at every turn, but it comes down to this: If race is not mentioned, in the absence of other signs (for example their name), we assume they are white; if gender is not mentioned, we assume they are male; if sexuality is not mentioned, we assume they are heterosexual; if ability is not mentioned, we assume no disabilities; and so on.

It should now be clear why, in erasing markers of identity difference online, we are not doing away with identity at all. We are merely resetting the cultural defaults. So if I am chatting online with a stranger named Cybernaut (an appropriately techie but generally neutral name), what mental picture will I form of that person, and how will I treat that person, perhaps subconsciously, because of this mental picture? Beth Kolko, Lisa Nakamura, and Gilbert Rodman summarize the situation this way:

> To be sure, as many white people point out when faced with questions of racial politics, race shouldn't matter. While we sympathize with the noble belief in egalitarian tolerance at the heart of such a response, we also recognize that the way the world *should* work and the way the world *does* work are two very different things—and that we live in a world that doesn't come anywhere close to that ideal. Whether we like it or not, in the real world, race *does* matter a great deal....
>
> Moreover, in spite of popular utopian rhetoric to the contrary, we believe that race matters no less in cyberspace than it does "IRL" (in real life). One of the most basic reasons for this is that the binary opposition between cyberspace and "the real world" is not nearly as sharp and clean as it's often made out to be. While the mediated nature of cyberspace renders invisible many (and, in some instances, all) of the visual and aural cues that serve to mark people's identities IRL, that invisibility doesn't carry back over into "the real world" in ways that allow people to log in and simply shrug off a lifetime of experiencing the world from specific identity-related perspectives. You may be able to go online and not have anyone know your race or gender—you may even be able to take cyberspace's potential for anonymity a step further and masquerade as a race or gender that doesn't reflect the real, offline you—but neither the invisibility nor the mutability of online identity make it possible for you to escape your "real world" identity completely. Consequently, race matters in cyberspace precisely because all of us who spend time online are already shaped by the way in which race matters offline, and we can't help but bring our own knowledge, experiences, and values with us when we log on.[13]

Indeed, as Nakamura demonstrates, real-life racial stereotypes crop up frequently in the online personae created for raceless cyberspace.[14] While someone proclaiming his or her Asian identity will be verbally attacked as being inappropriate, others are free to wander around dressed up as virtual geishas and samurai.

Online environments allow us the opportunity to challenge identity, but can only do so by recognizing how deeply embedded our cultural allegiances are, both in terms of our identities and our assumptions about others. Ultimately identity can be challenged online only by talking openly about it (online and offline), not by denying its existence and silencing those who would talk about it.

Cyborg Identity

Beyond challenging typical markers of identity such as gender and race, technologies participate in challenging the idea of being human itself. In this section we consider the way that the *cyborg*, the space where humans and technologies meld, challenges the very identity of the human body. In the next section, we consider the way that artificial intelligence challenges the identity of the human mind.

Technology challenges our understood notion of what it means to be human by transgressing the boundaries between our bodies and our tools. A *cyborg* (short for cybernetic organism) is an entity part human and part machine. In a rather renowned essay, Donna Haraway argued that such couplings and combinations of humans and technologies are potentially politically progressive because, in refusing to be just technology or just human, the cyborg rejects the cultural dichotomy between technology and human.[15] When asked, "What are you, machine or man?" the cyborg states, "Both and neither." Haraway argues that in refusing to choose, the cyborg acts as an ironic political model for challenging other divisions of identity like race, gender, sexuality, and so on, thus overcoming the technologies of categorization we have been discussing. She makes "an argument for *pleasure* in the confusion of boundaries and for *responsibility* in their construction."[16]

A body of literature has built up around this notion of the cyborg since Haraway's article was published in 1985. Some people claim that we have become cyborg, given the increase in artificial limbs, organs, life support, and life-enhancing devices implanted in humans; the increasing tendency for humans to communicate through technologies such as computer networks and cell phones; and the development of advanced robotics and automation technologies. Cyborgs like these are most often seen in science fiction. *The Six Million Dollar Man*, a television show in the 1970s, featured a cyborg, an astronaut severely injured in a crash who is saved by replacing some of his body parts (legs, an arm, an eye) with mechanical versions. Cyborgs readily march across our theater screens in films such as *Blade Runner*, *AI*, the *Alien* films, the *Terminator* films, the *Robocop* films, the *Matrix* films, and even the *Star Wars* films. They are also featured prominently as our future selves in such nonfiction works as PBS's documentary, *Beyond Human*.[17] But in many ways most of us are already cyborgs of one sort or another. Many people have artificial hips, some have artificial hearts or heart valves, and some are periodically hooked up to dialysis machines that filter their blood. But on an even more banal level, many wear glasses or contact lenses, and there are few people who haven't been subject to the technology of inoculation. To be a cyborg is not something new. Indeed, arguing from the logic of articulation and assemblage, which insists that we consist of a range of connections to language, technology, bodies, practices, and affects, we have always, a sense, been cyborg.[18] However, the cyborg, as a contemporary political argument, occupies a unique conjunctural position from which to challenge the received view presented in Section I of this book.

The idea of the cyborg is often used to highlight the new ways we are becoming cyborg through advanced and evermore-intrusive technologies. As an increasing number of bodily functions and organs are supplemented and even replaced by machines, concern grows that we might go too far, lose our humanity, and become mere machines. We begin to ask, what, after all, is human? At what point does someone become "more machine now than man" as Obi-Wan Kenobi says

of Darth Vader in *Star Wars*? As this line of questioning continues, the usual answer is that what makes humans human is not a certain combination of organs and
limbs, but our minds. Even if our bodies are destroyed—our minds left as brains in
a vat or downloaded into a computer or a robot—we feel that there is something
human that remains. The film *Robocop* is essentially about this theme. In the end,
the movie suggests, the human part of the cyborg will always prevail.

Artificial Intelligence (AI)

What if a completely artificial mind is developed that matches or exceeds the
capabilities of a human mind? What is so special about humans then? When our
intellect is surpassed, what is left to make us human? Emotion perhaps? The assertion that machines are mechanical and not emotional is another frequent theme
of science fiction films. But what if artificial intelligences can match human emotions? This is, after all, the goal of Rosalind Picard's research, discussed in Chapter
12. What then?

The project of artificial intelligence aims to produce an autonomous computer that "thinks" like a human, is aware, can learn, and can evolve. There have been
many tests devised to determine if computers "think." Interestingly, however, they
all rely ultimately on shifting senses of what it means "to think" and "to be."
Depending on how one designs the tests and defines "thinking" or "living," computers can be seen as thinking, or even living. But more to the point, what is happening is that the sense of thinking and being is changing in relation to the changing technological assemblage. This can be illustrated by examining one measure
of the intelligence of computing: the Turing Test, developed by one of the early
computer pioneers, Alan Turing.[19] Turing postulated that a computer would be
deemed intelligent if a human could hold an extended conversation with it and not
be aware that he or she was conversing with a machine. The test is usually set up
so that the observer or judge has a series of conversations via a computer terminal
with several people at terminals in another room. However, one or more of the
"people" with whom the judge is having a conversation will actually be a machine.
If the judge cannot tell the difference between the humans and the machine, then
the machine is said to have achieved a certain level of intelligence.

Journalist Charles Platt got the opportunity to participate in a Turing Test by
being part of the annual Loebner contest in 1995. Hugh Loebner offers $100,000
to the first person who can program a computer to fool ten judges according to
Turing's criteria. Platt was picked, not as a judge, but as one of the humans responding to the judge's questions. All he had to do was respond "naturally" to a
series of questions on a topic he had chosen (cryonics). The question occurred to
him, what happens if the judge thinks he is a machine? Or if one of the machines
is deemed "more human" than he is? How should he answer the questions in a
"human" way? If one is polite and carefully correct, one may seem like a computer.
But one can also program a computer program to include typographical errors.
One could joke with the judge, but computers can be programmed to do that as

well. In the end, Platt was not mistaken for a machine; in fact, he was awarded the most human human award for just being himself (as he puts it, "moody, irritable, and obnoxious").[20]

The problem with such a test is that the judges know that one of the respondents is definitely a computer; they just have to identify the right one. In other online settings such as chat rooms, MUDs, and MOOs, the same locations where other sorts of identities are challenged, participants can easily be fooled by bots. Bots are computer programs that appear as characters and carry on conversations with other participants. Some people quickly realize they're talking to a machine, others are completely taken in.[21] The difference is that in the nonexperimental environment people aren't prepped to "detect the bot."

Sherry Turkle has pointed out that as children use computers, they tend to attribute human characteristics to the computers; and as computers become more sophisticated, they can mimic more human characteristics.[22] This does not mean that the children are treating the machines as if they were human; they know they are just machines. But what this does is begin to shift the children's own definitions of what human is. Furthermore, they take it for granted that a computer has a personality and can be smart, silly, witty, or grumpy. Children growing up with these kinds of permeable boundaries between machines and humans will grow up to be very different kinds of adults than those most of us still are.

Artificial Life (A-Life)

Notions of human identity (and uniqueness) are even further challenged by developments in the field of Artificial Life. According to Margaret Boden, "Artificial Life (A-Life) uses informational concepts and computer modeling to study life in general, and terrestrial life in particular."[23] It seeks to understand the processes of life and to try to reproduce these processes, or lifelike behaviors, through computer modeling and robotics. A-Life's range of applications runs from realistic computer animations to pharmaceutical research and computer art. Discussions of A-Life usually involve questions of chaos theory, self-organization, and emergence. It is obviously centrally involved with questions of what "life" is and the possibilities for artificially creating new life. Christopher Langton, a key figure in the field of A-Life, describes one aspect of the project of A-Life as follows:

> Artificial life involves attempts to (1) synthesize the process of evolution (2) in computers, and (3) will be interested in whatever emerges from the process, even if the results have no analogues in the "natural" world.[24]

The A-Life approach to technology is to create technologies that react to and learn from their environment, therefore developing and changing as the result of their experiences of the world (whether that world is the memory of a computer or out among other robots and actors in physical spaces). If a robot, for example, moves through and reacts to its environment not in a preprogrammed way but in a way it has learned from its previous experiences of this and other environments, to what extent can we say that the robot is *alive* versus saying that it is more *lifelike* than programmed robots?

Conclusion: Why Challenge Identity?

In this chapter, we have explored ways that technologies have challenged our received notions of cultural identity, with implications for the designation and treatment of people according to the potent classifications that circulate in technological culture. They have also challenged our assumptions about what it means to be human, and even what it means *to be*. Technologies, particularly medical technologies, have participated in altering our definitions of what it means to be alive or dead. Given the state of technological culture, we are now faced with the problem of defining someone as dead when we can keep their body functioning with machines. Where, given the new medical technologies, is the (new) crucial border point between life and death? Beyond questions of life and death, new scientific technologies raise similar sorts of questions about the border point between life and nonlife. Technologies allow us to find life where we thought none would be, as in recent determinations that chains of proteins are "alive." Nowhere is the distinction between life and nonlife more politically charged than in the issue of the status of fetus and mother.

Technologies of monitoring fetuses in pregnant women (electronic monitoring of fetal heartbeats or ultrasound imaging) have helped to change the status of both fetus and mother. Presenting tangible evidence of fetal life (Here's the heartbeat! Here's the 3-D picture!) creates a technological quickening of the fetus much earlier than felt fetal movements, and attributes subjectivity to the fetus much earlier as well. In other words, the fetus is more likely considered a person—from a medical standpoint, a patient, and from a political standpoint, a subject—long before the final trimester of pregnancy, long before the fetus is at a stage to survive outside the womb. As a consequence of these technologies, the pregnant woman herself can become more invisible, making it possible to restrict her rights in favor of the fetus's.[25]

How could it be any clearer that technologies are not mere tools that we take up to accomplish particular ends? How could it be any clearer that technologies are not mere causes that have effects on what we do? In fact, the very (changing!) idea of who we are, how we think, and how we act is *articulated within, caught up in* a changing technological culture.

CHAPTER FIFTEEN

Politics

Technology Is Political

IN SECTION III OF THIS BOOK we have been exploring ways that non-necessary technological relationships (articulations, assemblages) affect the cultural and spatial definitions, arrangements, and distribution of peoples and technologies. There is a simpler way of stating this: Technology is political.

When we say that technology is political, we mean much more than to simply say that technology has political uses or that it can be mobilized for political ends; although this is surely part of a technology's politics. And by "politics" we mean something more than the sorts of politics encountered watching television news: the politics of government, political parties, political conventions, and elections. To explain what a technological politics looks like, we first clarify what we mean by politics, then how technology is political. Along the way we point out why so many people think that technology is not political. Our final point, and indeed this is one of the central motivations for writing this book, is to answer the question: What does the insistence that technology is political and the attention to a technological politics generally get us? The short answer is that it feeds the desire for greater participation in the cultural processes that shape our lives.

What Is Politics?

Political scientist Langdon Winner has been very influential in promoting the notion that technologies have politics. His definition of politics is an instructive place to begin. Winner writes: "By the term 'politics' I mean arrangements of power and authority in human associations as well as the activities that take place within those arrangements."[1] There are two aspects to this definition of politics: First, it concerns human associations, or, relations among people. Second, these relations are relations of power. We will discuss each of these parts in turn.

First, to address the nature of human association referred to in this definition, we note that human associations are typically understood to be social, in that they constitute one's sociological position in society and arise out of such positions. Thus, factors such as one's social status, race, and income shape how one associates with others. However, a cultural studies perspective insists that human associations are cultural as well as social. By this we mean that social categories such as status, income, and race are not fixed realities, the meanings of which are obvious and universal. Rather, as we discussed in the preceding chapters on identity, they are contingent identities that take on meaning and significance only within concrete assemblages. Thus, human association, as the content of politics, must necessarily include the ways that concrete assemblages entail cultural matters such as the creation of meaning, the configuration of space and time, the workings of identity, the sense of possibility, and the affective "feel" of everyday associations.

In addition, as we have highlighted throughout this book, those everyday human associations are also always technological. Even at the most rudimentary level, our associations with others involve the technology of language. Beyond that, all manner of technologies are involved in these associations: from telephones and automobiles to reproductive and genetic technologies. If politics has to do with human associations, we conclude that it has to consider technological associations as well. As another political scientist, Richard Sclove, has put it, "technologies also represent an important kind of social structure. By 'social structure' I mean the background features that help define and regulate social life."[2] But here too, by insisting on the cultural, we expand and deepen what Sclove means by the social.

Second, to address the aspect of power in this definition of politics, we have pointed out in previous chapters that the relations (or articulations) between humans, between humans and technologies, and between technologies and technologies are not fixed, but contingent. These relations are made and unmade, and they are made to mean. A politics of technology is about the making and unmaking of these connections, the arrangement of actors, and the articulation of possibilities. These contingent articulations (connections, arrangements) always have to do with power, the ability or inability to achieve effects, and they always involve relations of agency. What we mean by politics is the power to articulate, to make arrangements of peoples, technologies, and languages, to make things happen or not, to give an assemblage its shape. Therefore, it makes sense to think of technology as political.

Why Not Politics?

Not everyone would readily agree that technology is political, which is why we briefly turn to the question: Why do people think technology is not political? We identify at least three reasons. The first is their view of technology itself; the second is their definition of politics; and the third is a resignation to the status quo.

First, as we argued early in this book, the dominant view of technology claims that it is a neutral tool. This is the NRA's "guns don't kill people, people do" argument—if technology is a neutral instrument, it certainly can't be political in and of itself. As Winner puts it, "Blaming the hardware appears even more foolish than blaming the victims when it comes to judging conditions of public life."[3] Even though one can usually get people to accept that a technology can be put to political uses (for example, using closed-circuit televisual surveillance systems to control the behavior of people in a public place), few people will accept that the technology (the camera, in this case) is political.

The second reason people have difficulty accepting that technology is political has to do with the standard definition of politics. As we have pointed out, the widely circulating definition of politics is usually limited to institutions and issues of governance, and it is usually restricted to describing people. Prevailing definitions of politics, like prevailing definitions of society, exclude the nonhuman. For example, in saying that people kill people in the NRA statement about guns, the technology disappears entirely from the question of what is doing the killing.

The third reason people have difficulty accepting that technology is political is more complex, but ultimately quite important. Most people tend to accept the status quo because they have not been taught or encouraged to look beyond the current order of things, which is deeply built into the institutions and structures of everyday life. If Sclove is right, and technologies are a kind of "social structure" in themselves, then our technologies are deeply implicated in supporting the status quo and in providing a conservative sense of stability in life.

However, despite such popular phrases as "politics as usual," which equate politics negatively with maintaining the status quo and reduce the idea of politics to governance, the promise and practice of politics carry with them the basic presupposition that change is possible. To raise the question of politics thus opens the door to possibilities for change, which threatens a conservative commitment to the status quo. To raise the question of the politics of technology opens the door to challenge the complex assemblage within which much about technology and culture might be changed, which threatens the institutions and structures of everyday life. This is perhaps why people resist doing so, and it is precisely the point of doing so here.

What a Technological Politics Looks Like

The things we call "technologies" are ways of building order in our world. Many technical devices and systems important in everyday life contain possibilities for many different ways of ordering human activity. Consciously or unconsciously, deliberately or inadvertently, societies choose structures for technologies that influence how people are going to work, communicate, travel, consume, and so forth over a very long time.... Because choices tend to become strongly fixed in material equipment, economic investment, and social habit, the original flexibility vanishes for all practical purposes once the initial commitments are made. In that sense technological innovations are similar to legislative acts or political foundings that establish a framework for public order that will endure over many generations.[4]

The relationship Winner describes here among technologies and law, legislative acts, and political institutions, is not metaphorical. They are, as he says, similar, which is of considerable significance. The founding documents of the United States—the Constitution and the Bill of Rights—were crafted in processes of extensive deliberation. Laws passed by Congress are, for the most part, deliberated. But technological acts, though they often have equal or greater impact on institutional structure and culture than laws do, are rarely deliberated. Instead, we seem to proceed in a state of "technological somnambulism," as Winner calls it, sleepwalking through sweeping technological changes in everyday life.[5] Why is this so?

One reason for failing to attend to technological choice as political choice is, once again, the power of the assumption that technologies are neutral instruments. We live the paradox: On the one hand, we defer to the experts whose domain is technology—scientists, engineers, technicians, and, increasingly, business interests—as though they will make technology work on our behalf. On the other hand, we acknowledge that technology can be abused, but do not consider it our business to intervene in the process. Once again, an ingrained commitment to progress at any cost can blind our ability to see that we might have something to say or do about how technological culture develops.

A second reason that technology is not truly deliberated has to do with the theoretical tools and language available to the people and groups that do occasionally argue against some aspect of the technological assemblage. Without a thoughtful alternative to the received view, they commonly rely on theory and language that, despite their best efforts, does little more than support the system as it is. Winner, in *The Whale and the Reactor*, examines concepts and political language that have been used to engage a technological politics that has not been particularly helpful: efficiency, decentralization, revolution, nature, risk, and values.[6] Each has implications that deserve further scrutiny, because when debates over the place of technology are posed in these terms, they function as traps that keep debate safely ensconced within the status quo. A critique of the use of these terms reveals the criteria that are used to make technological decisions and, crucially, the criteria that are discounted. Below we highlight the work of three of these terms: efficiency, revolution, and nature.

Winner explains that arguments about efficiency reduce questions concerning technology to what is easily quantified: inputs/outputs and cost/benefit. Further, the attention to efficiency relies on a value of pragmatic necessity; technologies are practical matters that can only be discussed in practical terms. Who, after all, in North American technological culture is likely to argue that less efficiency might be desirable? That wouldn't be practical. Yet, as a result of attention to efficiency, debates over justice, humanity, and morals have almost no place. While we may care if a technology is more environmentally friendly or more just, the discussion almost always comes back to the bottom line: Is it efficient? Will it cost more? Even movements like the Appropriate Technology movement of the 1970s got caught up in the language of efficiency, when, for example, it was typical in the pages of the *Whole Earth Catalog* to encounter the focus on the efficiency of some new technology (like solar heaters), rather than how it served goals such as dignity, justice, equality, reciprocity, and creativity.

Winner also points out that the language of revolution—the talk that permeates the advent of the computer, the Internet, biotechnology, and nanotechnology—does not ask the question usually posed about political revolutions: What are the goals of the revolution?[7] As Winner argues, technological means have supplanted human ends: The means give shape to the ends, rather than the ends giving shape to the means. Rather than deliberating over goals and seeking the means to achieve them, we generate lots of new technologies (the means) and hunt for goals or ends to justify and/or sell them.

As Jennifer has written elsewhere, revolution in the technological context is used to indicate "fundamental change," in which everything is overturned, which is considered progress, and which is taken to be brought about solely by the technology. Thus, built into the concept of technological revolution is the notion that it is good and that it is out of our control.[8] Having argued against both of these positions in this book, we submit that a technological politics needs to be able to deliberate and ask: what are our goals? What technologies (if any) will help us get there? How could they be implicated in accordance with those goals?

Even the appeal to nature must be critiqued; for, as Winner writes, "Nature will justify anything"[9] Indeed, as Jennifer's research has demonstrated, appeals to nature can be used to make all manner of political arguments.[10] It matters if nature is something that humans have an inalienable right to exploit and use, if it is something to be preserved, or if it is a source of lessons and morals. Very different forms of politics emerge from these different views. The concept of nature is further complicated by its relation to the human everyday, that is, the site where the political realities within which matters such as use, exploitation, preservation, and conservation take on real meaning and significance. Thus, the lesson is to refuse to accept appeals to nature at face value, but to craft carefully a thoughtful sense of the place of nature in technological culture.[11]

Ironically, even though North American technological culture has not been crafted in deliberation and debate, there is what Winner refers to as a "defacto... sociotechnical constitution...of sorts" in place.[12] This constitution, which gives meaning and shape to cultural relations, has been developed by economic and ideological interests, is embedded in structures, institutions, and practices, is the ongoing production of political struggles, and is sustained by our acceptance of a particular political language and value structure. This constitution exhibits five interrelated characteristics, according to Winner:

1. The "ability of technologies of transportation and communication to facilitate control over events from a single center or small number of centers."[13]

2. The "tendency of new devices and techniques to increase the most efficient or effective size of organized human associations," which leads to gigantic centralized corporations and organizations.[14]

3. The tendency to "produce its own distinctive form of hierarchical authority," which in the workplace is "undisguisedly authoritarian."[15]

4. The tendency of "large, centralized, hierarchically arranged sociotechnical entities to crowd out and eliminate other varieties of human activity."[16]

5. The ability of "large sociotechnical organizations [to] exercise power to control the social and political influences that ostensibly control them. Human needs, markets, and political institutions that might regulate technology-based systems are often subject to manipulation by those very systems."[17]

Winner's comparison of our legal constitution and our sociotechnical one should be taken seriously. A technological politics ought to debate over new technologies in much the way that we debate new laws.

In spring 2003, Winner was asked to make a presentation before the Committee on Science of the US House of Representatives on the Societal Implications of Nanotechnology. In his talk, he advocated yet another specific parallel between accepted political practice and potential technological practice: the creation of juries of citizens. Juries are convened to establish a fellow citizen's guilt or innocence and to set the limits of punishment or reparation. Why couldn't a jury of citizens be established to hear cases about new technologies and to likewise rule? "The panels would study relevant documents, hear expert testimony from those doing the research, listen to arguments about technical applications and consequences presented by various sides, deliberate on their findings, and write reports offering policy advice."[18]

In many ways, Richard Sclove, founder of the Loka Institute, a nonprofit organization advocating democratic technologies, follows Winner's lead in examining the political process of technological decisions. Sclove, introduced in Chapter 13, is committed to the idea of *strong democracy*, which he borrows from political scientist Benjamin Barber.[19] Strong democracy advocates that citizens should have a role in making decisions over matters that affect their well-being and way of life.[20] It is an argument for more active citizen involvement in a whole range of civic matters in order to counter the perceived passivity of US citizens, who continue to vote less in general elections and tend to let elected representatives make most of the decisions. Sclove argues, like Winner, that since technology is a form of social structure that affects people's lives, citizens should have full democratic input into decision-making processes that involve them. He feels that those who use technologies and those affected by them might have more important insights into their functions, uses, and problems than technocrats, bureaucrats, or experts from outside the local context.

Rather than simply stating all this as an impassioned, though general argument, Sclove has described a list of nine design criteria, which specify values and processes for more democratic decision making with regards to technology. He says that these design criteria are only preliminary. He expects that they will be revised as they are used by local communities making their own decisions about technopolitical matters. Here are his criteria:

Toward DEMOCRATIC COMMUNITY:
A.   Seek a balance among communitarian/cooperative, individualized, and transcommunity technologies. Avoid technologies that establish authoritarian social relations.

Toward DEMOCRATIC WORK:
B.   Seek a diverse array of flexibly schedulable, self-actualizing technological practices. Avoid meaningless, debilitating, or otherwise autonomy-impairing technological practices.

Toward DEMOCRATIC POLITICS:
C.   Avoid technologies that promote ideologically distorted or impoverished beliefs.
D.   Seek technologies that can enable disadvantaged individuals and groups to participate fully in social, economic, and political life. Avoid technologies that support illegitimately hierarchical power relations between groups, organizations, or polities.

To help secure democratic self-governance:
E.   Keep potentially adverse consequences (e.g., environmental or social harms) within the boundaries of local political jurisdictions.
F.   Seek relative local economic self-reliance. Avoid technologies that promote dependency and loss of local autonomy.
G.   Seek technologies (including architecture of public space) compatible with globally aware, egalitarian political decentralization and federation.

To help perpetuate democratic social structures:
H.   Seek ecological sustainability.
I.   Seek "local" technological flexibility and "global" technological pluralism.[20]

We present these criteria as items for discussion, as an example of actualizing the sorts of assumptions that underlie Winner's arguments. No one else has so explicitly mapped out a set of design criteria based on the assumption of the political nature of technology.

However, the perspective of articulation and assemblage reveals some of the weaknesses of Sclove's approach. In particular, Sclove bases his philosophical and political position on the belief that the autonomous individual citizen is the heart of a democratic society. He doesn't question how the idea of the autonomous individual citizen may actually be a particularly tenacious articulation created during the Enlightenment and the early days of capitalism. Indeed, the same notion of the individual on which he bases his arguments also serves as the foundation of the technological cultural he critiques. Further, Sclove falls too easily into the trap of talking about technology as if it were an isolatable thing. Although he does acknowledge that technologies are contextual, he does not recognize this to the extent that articulation and assemblage allow us to. Thus, he misses the full significance of the need to critique the assemblage, not just a political process.

To understand the politics of technological culture, we need to take the ideas of articulation and assemblage seriously. As Jennifer has written, articulation is not just an exercise in epistemology; in other words, it is not simply another way of knowing and understanding technological culture. It is also not simply an exercise in theory or the generation of an interesting abstraction from real circumstances. Nor is it simply a method to describe how "best" to study technological culture. Articulation is about politics and strategizing.[22] Part of the task of articulation is to revisit assumptions—the very categories we take for granted, such as culture, technology, individuality, and democracy—to discover how these categories (their practices, representations, and affects) are historically constructed and articulated so that we can open them to scrutiny. When we scrutinize assumptions, we realize they are actually non-necessary contingencies. Part of the task of cultural studies is to discover how these contingencies are grounded in the specificities of the times and places wherein they occur. When we do this, we are taking context seriously, dealing with concrete historical conjunctures rather than generalities. Even Sclove argues that we cannot say that all windmills are democratic; the democratic capabilities of a windmill depend on the richly textured assemblage in which it occurs.[23]

Conclusion: Why Politics?

The point of this chapter has not been to introduce politics to the discussion of technology, but to insist that technology is political through and through. We would like to emphasize an observation with which we opened the book: how the received view of technology serves a particular political purpose. Arnold Pacey puts it this way:

> When people think that the development of technology follows a smooth path of advance predetermined by the logic of science and technique, they are more willing to accept the advice of "experts" and less likely to expect public participation in decisions about technology policy.[24]

The way we think about technologies affects what we think we can do about them. It makes sense to be passive if we think with the received view, for creative action is not supported by an unreflective commitment to progress, a goal of increased convenience, or a belief in technological determinism. In contrast, a cultural studies approach is capable of supporting creative intervention in the service of rearticulating the technological culture, by working through the critique of technological culture as it is manifested in everyday life. In doing so, it is important that we recognize the tenacity of the articulations that we tackle; the struggle to articulate and rearticulate technological culture is a long and involved one.

Antonio Gramsci describes two types of warfare fought on the political plane, which can be applied to struggles with and within the technological assemblage: the *war of movement or maneuver* and the *war of position*.[25] The war of maneuver refers to the concentration of forces in an all-out assault on one front that promises to breach the enemy's defenses and achieve a quick and complete victory. Rarely does struggle in the realm of the technological assemblage work this way. Rather, the war of position is the rule. A war of position takes place across many fronts, at many sites of struggle and resistance, and no battle is decisive. It is a slow, continuous struggle, with, at times, little movement to show for the struggle. Given the complex nature of the technological assemblage, made up of multiple, sometimes-contradictory contingencies, it makes sense that change is more likely to take place in this way. What is required is a *reterritorialization* of a complex assemblage: All manner of interventions will need to occur at all kinds of levels in all kinds of situations. Success in a war of maneuver may come slowly, if at all, but what is ultimately at stake is a reconfiguration of a culture.

Our struggle within technological culture is that of a war of position, of small victories, of the slow reterritorialization of discourse, the gradual rearticulation of objects, practices, representations, and affects in which every move, no matter how small, is an important matter. Part of this struggle is the call for new narratives and new stories about technological culture. Anne Balsamo described the project of feminist cultural studies in terms that apply equally here. The project, she writes, "is to write the stories and tell the tales that will connect seemingly isolated moments of discourse—histories and effects—into a narrative that helps us make sense of the transformations as they emerge." New stories, she continues, "also serve as expressive resources that offer cognitive maps of emergent cultural arrangements."[26] Rather than distanced analyses, such stories are *engagements*; they provide tools, frameworks, and maps for others to take up, critique, or to use as springboards for new narratives, actions, and politics.

In many ways we have framed the cultural studies answer to the problem of technology in terms of method. In other words, we are focusing on dimensions with which we may take a new perspective on the relations of technological culture. What we have not done is to articulate alternative criteria by which to judge, fashion, or transform technological culture. Such criteria circulate elsewhere and range from the more general appeals to justice and dignity to more elaborate schemas based on Buddhist principles, or principles of strong democracy.[27] It has not been our goal to establish such criteria or impose them. This is partly out of respect for the radical contextuality of our approach, and partly a desire to not impose new universals in place of the old. This is not to say that we have no preferences of our own. It would be disingenuous of us to claim political neutrality. Therefore, we wish to affirm the need for strong public participation in shaping culture, while recognizing the many ways in which we are already doing this in our everyday lives. We wish to affirm the need to subscribe to the values of justice, dignity, creativity, equality, reciprocity, democracy, and reverence for the environment in making technological and cultural choices.

We also need to acknowledge our own position within a dynamic global space. Other global spaces may present other possibilities, other stories, and other tools. We begin to address these other contexts in the next chapter.

CHAPTER SIXTEEN

Globalization

The Story So Far, and Other Stories

IT IS SPRING IN NORTHERN INDIA. In workshops and factories, decorations are going up, but not just on the walls. Machines and tools, including computers and fax machines, are decorated and prepared for the festivities. The shop floor is rearranged, even cleared, to allow for the placement of the deity. Outside the factories, cars and trucks are festooned with flowers and ornaments, and kites are flying on an early-autumn breeze.

This is the celebration of Vishwakarma Puja. Lord Vishwakarma is called "Lord of Tools," "God of Engineering and Architecture," and "the Lord of the Arts." In the Hindu religion, Vishwakarma is the creator (the architect) of the Universe, who, among his other feats, built great cities and designed chariots and missiles for the Gods. Every September 17, artists, craftsmen, architects, and all those who work and create with tools worship him. They pray that their machinery will serve them well in the coming year; they pray for inspiration for creative acts, and they resolve to work harder in the future.

In the holy city of Puri, on the coast of the Bay of Bengal, in late June or early July, another celebration takes place called Ratha-yâtrâ. This festival honors Jagannâth, the Lord of the World. Jagannâth is a form of the Hindu god Krishna, and the Jagannâth temple is one of the holiest sites to Hindus. During Ratha-yâtrâ the image of Jagannâth is placed in a gigantic vehicle, four stories tall. The vehicle is so heavy that it takes hundreds of worshipers to drag it slowly through the streets. It is said that some of the faithful used to throw themselves under the wheels of the vehicle in an act of devotion to the deity.[1]

The English word "juggernaut" derives from European perceptions and representations of this festival. *The Oxford English Dictionary* gives a definition of juggernaut as "an institution, practice, or notion to which persons blindly devote themselves, or are ruthlessly sacrificed."[2] The juggernaut is often considered a crushing vision of technological progress. But what a contrast between the juggernaut and the painting *American Progress* described in Chapter 1, in which beautiful liberty floats across the technological landscape bathed in the light of the rising sun. Despite the rich symbolism of the term, we need to make it clear that English and North American notions of the juggernaut have no necessary relation to Jagannâth or Ratha-yâtrâ. In terms of technological culture, Ratha-yâtrâ is more symbolic of the majesty and scale of technology and technological creation, which make it closer to something like the sublime.

We present the opening stories of this chapter not as quaint, "primitive," or exotic alternatives to contemporary North American technological culture. Rather, we mean to illustrate ways that different assemblages are configured in different contexts. Religion is in the foreground in the above examples, unlike in the North American received view of technological culture. Religion is not absent from the received view—far from it. Where, after all, would the received view be without Protestant notions of the work ethic and Manifest Destiny? However, religion is not explicitly acknowledged in the received view. It is articulated differently. Even the idea of what religion is, and how and where it is practiced, is articulated differently.

The Hindu festivals of Vishwakarma Puja and Ratha-yâtrâ are part of particular assemblages, but they alone (and Hinduism alone) do not account for the technological cultures of India. In the first place, the specificities of religious practice in India vary from site to site and from community to community. In addition, religious practices vary over time due to changing cultural and political circumstances. For example, British colonialism outlawed some practices and allowed others to continue. Further, the current political dominance (or hegemony) of a conservative Hindu political formation empowers some practices and disinvests others (especially those that are not Hindu, of which there are many in India). But even with these caveats in place, is this enough to grasp the assemblages of technological culture in India? Not even close. We would also have to consider the legacies of colonialism, including the place of the British-built railways and other imported technology. We would have to consider the legacy of the Indian revolution, led by Mohandas K. Gandhi, and its symbolic use of another technology: the spinning wheel, which now appears on the Indian flag. We would have to consider the ongoing drive for nationalist development led by India's first prime minister, Jawaharlal Nehru, in the late 1940s and 1950s, which was based on ideas of technological development derived from the received view of technological culture in the West.[3] Large-scale technological projects like dams, electrification schemes, satellite systems, nuclear programs, and agricultural programs such as the "the Green Revolution" have all represented progress as a national ideal at various conjunctures. We would have to consider the incredible disparities in income, health, lifestyle, and literacy within the Indian population, and the disparities in political outlook and organization from state to state. We would have to consider the vast pirate subcultures scattered throughout India's urban areas: During the 1990s most cable television systems were illegal, pirated affairs. Today an ad hoc pirate cyberculture exists that cobbles together computers from spare parts bought in crowded market stalls, and establishes computer networks with old BBS software and recycled modems.[4] We would have to consider the great number of other religions, languages, and cultures across India and the subcontinent, and the global population of Indian technicians, scientists, medical providers, taxi drivers, and others working in the Middle East, Asia, and North America.

We could go on, but the point is not to describe the technological culture of India in its entirety, but to illustrate that there are other assemblages around the globe, and that the ones found in North America are particular in their context. Japan, Russia, Indonesia, Australia, and Nigeria all have their particular assemblages of technological culture. However, these assemblages are not entirely distinct, in that assemblages overlap. For example, we might note the presence of Western notions of nation, progress, and development in the story about India above, or consider the presence of ideas of technological modernism in Japan, Russia, Indonesia, Australia, and Nigeria. Such overlapping notions become more common, but no less complex and particular, in a global age.

Globalization

Paul du Gay has expressed the importance of the concept of globalization in this way:

> "Globalization" has become possibly the most fashionable concept in the social sciences, a core axiom in the prescription of management consultants, and a central element of contemporary political debate. The concept of "globalization" has achieved such widespread exposure, and has become such a powerful explanatory device and guide to action, that it sometimes appears almost unquestionable. Certainly its effects have been pronounced.[5]

The fact that globalization is unquestionably accepted as an ongoing process by both its proponents and opponents should be an indication that the term is in need of deconstruction; we need to discover the specific articulations that make up globalization as it is generally understood. This is an issue that currently calls for penetrating cultural analysis. We merely raise the issue here to flag the fact that we do not accept the term at its face value, but that we recognize the specificity and historicity of the processes it indicates. What has been called globalization has been driven very much by the ideas of progress, convenience, technological determinism, the technological fix, and the desire for control. A key component of the concept of globalization is the way that these components of the received view of technological culture are condensed in the theory and practice of *development*. Development entails the export of technologies, technological systems, technological products, and scientific knowledge from "developed" Western countries to the "underdeveloped" or "developing" world, in order to help them industrialize and modernize. Those advocating development often ignore the specific contexts of the reception of the products and knowledge, ignore local conditions, and ignore local knowledge. For example, as part of its nationalist development scheme mentioned above, India created a project in the 1970s to send agricultural information, development programs, and educational programs for children to select states across the country in the form of television programs transmitted via a NASA satellite. While the SITE project (Satellite Instructional Technology Experiment Project of 1975–1976) was a considered a technical success, debates still question whether or not it was a cultural success. Receivers were constructed in villages, programs were transmitted, and people were able to view them across the country; but how could one consider it success, for example, to transmit agricultural information pertinent in one region to another?[6]

Crude versions of globalization describe the process as a steady expansion of products and ideas from the "West to the Rest," in a way of inferring a process of the West colonizing the rest of the world. But it is crucial to note that globalization actually consists of multiple flows in multiple directions and dimensions, some even from the "Rest to the West." Arjun Appadurai describes globalization as a series of landscapes, each with its own independent movements.[7] He points out that what is most important about these landscapes are the contradictions or disjunctures between them. Because they continually contradict each other in different ways, he argues, globalization will never become a process of homogenization. Appadurai identifies five landscapes (more simply, *scapes*), what he calls "five dimensions of global cultural flows":

1. *Ethnoscapes*: the movement and placement of people (and peoples), the increasingly mobile landscape of immigrants, refugees, tourists, workers, and others. People as they move bring with them culture, language, things, ideas.

2. *Mediascapes*: the movement, production, and display of mediated images and information, from Hollywood films to CNN to regional media centers like Brazil, Hong Kong, and India, and alternative networks like Indymedia.

3. *Technoscapes*: the movement of technology, especially the investment in industrial technologies and factories globally, but also the movement of other technologies (used in a more traditional sense of the term).

4. *Finanscapes*: the movement of money, of global capital, both in stock and currency markets and in World Bank loans and other financial investment.

5. *Ideoscapes*: the movement of political ideas like democracy, freedom, rights, and so on.[8]

Given the significance of these global flows, a consideration of technological culture in the global context must do more than simply consider how a particular assemblage has nonlocal components, such as offshore factories, foreign investments, guest workers, and foreign technologies. It cannot limit its purview to the flow of technologies (the landscape of the technoscape) alone. Instead, it must consider all the flows in relation to one another. If we were to map the technological culture of Finland, Eritrea, the Philippines, Uruguay, or Nepal, we would have to consider the placement, displacement, and movement of people, media, technology, money, and ideologies in order to begin to grasp the specificity of that particular conjuncture.

Antiglobalization

While some view the processes of globalization as inevitable, positive, creative, and profitable, others are much less sanguine and see in these developments the attempt of a handful of powerful countries and multinational corporations to dominate global markets and culture and to impose a value scheme based on competition, profit, and efficiency. Currently the most prominent place to find overt challenges to the received notion of culture and technology is in the multifaceted antiglobalization movement, which in many ways came into the public eye with the protests in Seattle against the meeting of the World Trade Organization in 1999. Rather than promoting a single cause, the movement is a loose conglomeration of a variety of groups advocating on behalf of a number of issues: the environment, fair trade, labor rights, women's rights, immigrant's rights, indigenous people's rights, food safety, anticonsumerism, anticapitalism, and many others. In the space that we have here, it would be impossible to do justice to all these movements, and speaking in general terms about commonalities between them is ultimately unfair to the diversity of positions and solutions offered by these groups. Let us at least provide a couple of examples.

Writing under the label of ecofeminism, Vandana Shiva has become an inspiration for some antiglobalization efforts. Shiva, a physicist, philosopher, and activist, argues more generally against the problems inherent in the model of development discussed above, particularly its negative impact on women.[9] More specifically, Shiva argues that the imposition of Western industrialized-agricultural techniques on the rest of the world is not only wrong but dangerous. It is dangerous because such techniques advocate developing agricultural monocultures (not only planting single species but single seed stocks, thus decreasing biodiversity). Such practices of monoagriculture leave the farmer dependent on extensive government-subsidized irrigation (since generally these techniques require much more water than indigenous crops), pesticides, chemical fertilizers (since the introduced species need help to survive within the new environmental context), and seed purchased from multinational agribusinesses (seed that does not regenerate itself, but needs to be purchased again the next year, rather than using locally produced, renewable seed). We can see in these practices the presence of the received view of technological culture: Chemicals, new farm machinery, and genetically modified seed must constitute progress; monoagriculture is harvested much more efficiently (by machine rather than hand); and the process of agriculture is seen as a problem that can be fixed technologically. Shiva goes on to point out how this same assemblage of corporations, science, and the technology of development add insult to injury by allowing the patenting, and therefore ownership, of particular varieties of seed, even if such varieties occur in nature. Indigenous peoples may find that they have suddenly lost the right to plant the seeds they have used for generations. Shiva writes:

> Laws for the protection of private property rights, especially as related to life forms, cannot and should not be imposed globally.... By keeping human rights at the centre of discourse and debate, we might be able to restrain the ultimate privatization of life itself.[10]

A number of mobilizations against globalization have taken up the issue of agricultural patents and indigenous rights by focusing on genetically modified (GM) organisms. These mobilizations have included public protests against the sale of GM food in Europe, the United States, and in other parts of the world, which in some places have resulted in the banning of GM food. In many ways, these actions are also against the global dominance of corporations and corporate culture's focus on profit and property. In an overview of anticorporate movements, Amory Starr includes pro-pedestrian and pro-bicycle movements against "the corporate project of automobile use;" anti–fast-food movements, especially anti-McDonald's movements; international boycotts against pesticide companies; and international boycotts against companies that employ child labor and operate sweatshops.[11]

What is interesting about the various antiglobalization movements is that, on the whole, they tend not to highlight technology (in the sense of artifact or device) as the problem, which is why we did not include them in Section II. They do have some affinity for the Appropriate Technology movement of the 1970s, but the success of these movements depends primarily on utilizing new high-tech technologies. In particular, their organization of people, resources, knowledge, and information relies on sophisticated technologies of communication. The dissemination of their information and position statements and the coordination of the regular and massive protests—which have disrupted global conferences and meetings since Seattle—are precisely dependent on new technologies for their success. The fact that these groups are more successful now than they have ever been owes much to their use of the Internet to connect with populations and groups that otherwise would have remained disconnected. The integration of cell phones and pagers has also been a decisive factor in organizing, particularly during protest events, where they are used to send voice and text messages to rapidly inform and redirect large groups of people in response to developments anywhere in the protest zone: The police are here; the television crews are there; these streets are being closed; or those people were just tear-gassed.

An example of a group that has successfully integrated new communication technologies is Mobilization for Global Justice, which, in conjunction with the Latin America Solidarity Coalition, organized protests against the annual World Bank and International Monetary Fund meetings held in Washington, DC, in

April 2003. Their Web sites (including lasolidarity.org, globalizethis.org, and a16. org) provide a detailed calendar of protest events; suggestions for protest signs and actions; information on childcare, housing, transportation, and medical care; and numerous other resources for activists, including strategies for dealing with the media, downloadable posters, and legal information in case of arrest. This sophisticated use of new technology is characteristic. The following is a description by Graham Meikle of how protesters used the Internet to organize the 1999 Seattle protests:

> Groups preparing for Seattle did much of their work on the Net. Established organizations—the Sierra Club, Corporate Watch, unions—were joined by new coalitions, such as the Direct Action Network, which grew from three groups to 70 in the months leading up to N30 [November 30, the date of the protests]. Websites offered resources to those planning to attend, from maps of Seattle to legal advice, while email and listservs were used to coordinate, inform, organize, and train. The Ruckus Society used the Net to provide manuals and organize action camps for training in nonviolent civil disobedience, from urban abseiling to crafting soundbites. And as well as influencing the offline events, cyberspace was also the actual site of some anti-WTO action. Corporate sabotage specialists Æ_ark (pronounced 'art-mark') created a sophisticated parody of the WTO's website to provide counter-spin, while the Electrohippies mounted a virtual sit-in of the real WTO site. And when events began, the online Independent Media Centre uploaded real-time reports from activists in Seattle, and streamed audio and video coverage.[12]

Antiglobalization groups such as these are actively engaged in a very visible war of position (that verges at moments on maneuver) to rearticulate the multiple components of contemporary technological culture. Whether one agrees with their goals or not, it is difficult to ignore the significance of their interventions.

Conclusion: Why Globalization?

The inclusion of the global as a dimension in the analysis of technological culture underscores at least three important insights. First, it jettisons us from the ethnocentric, parochial worldview that considers our own experience to be universally applicable. A corollary to the ethnocentric worldview is the belief that the rest of the world is merely peripheral to what "really matters," or, at best, an "exotic" amusement. The global reminds us that conjunctures are particular to times and places, and what "really matters" depends on a willingness to map the flows that constitute other ways of understanding other conjunctures. Second, the global reveals intersections of our conjunctures with the rest of the world (or parts of it) in which there are differences as well as commonalities among articulations and assemblages. Further, it reveals that local actions can have global consequences, and that global actions can have local consequences. Third, the global draws us necessarily into other places in the world. It does not permit us to ignore other people and places. It demands that we critically engage the workings of a complex *global* technological assemblage.

Miscellaneous Subjects: Telephone, Directory and Globe
Photograph by Theodor Horydczak, ca 1920–1950
Library of Congress
Horydczak Collection

Conclusion:
The Cultural Challenge

IN APRIL 2000, BILL JOY PUBLISHED the rather astonishing article "Why the Future Doesn't Need Us" in *Wired* magazine.[1] The article, which we have referred to previously in this book, argues that humankind itself is threatened by new technological developments, in particular new technologies of robotics, genetic engineering, and nanotechnology. For Joy, these new technologies are more dangerous than weapons of mass destruction, whether nuclear, biological, or chemical, because they can potentially self-replicate. A smart robot can build and program other robots; new pathogens can self-replicate and cross species barriers; and machines on the nanoscale could rapidly make millions more molecular machines in a process that we might be unable to stop. Such nightmare scenarios have been sketched before, but what made people sit up and take notice of this article is the pedigree of its author: Joy is a founder and chief scientist of Sun Microsystems, the creator of the Java programming language, and a well-respected member of the computer establishment. He is unaccustomed to questioning the consequences of his own work and those of his colleagues, yet in 2000, this is precisely what he did. There are those who disagree with Joy's analysis of the situation and his speculations about the future; they argue that he is too pessimistic; that the benefits far outweigh potential harms; and that the effects will not be as disastrous as he describes. Nonetheless, these technologies are, and will be, a challenging part of our technological culture.

Today's robotics are much different from attempts to visualize or build robots in the past. Rather than building self-contained, preprogrammed autonomous machines, contemporary robotic research seeks to build robots that will learn to interact with their environment and with other actors. A related goal is to produce robots that will function seamlessly as part of our daily environment by building human-like behaviors, reactions, responses, and appropriate emotions into the machines around us.

Whereas genetic change tends to favor adaptations to particular environmental circumstances, today's genetic engineering tends to favor features that have economic value: so argue Amory Lovins and L. Hunter Lovins in an article that accompanied Joy's.[2] According to them, the problem with genetic engineering, including GM foods, GM organisms, and the entire Human Genome Project, is that scientists *think* they know how genes work; but they might very well be wrong. For example, scientists think that specific gene sequences have specific effects—for example, that there is a gene for blue eyes—but they don't know for sure how gene sequences interact with other sequences. In fact, there is a lot of genetic matter for which the function is simply not understood; but scientists simply disregard that. Danger lies in the fact that we may not recognize possible negative consequences of such genetic rearrangements for decades. As Paul Virilio argues, we have created a genetic bomb: a situation in which a single accident could threaten all of human life by undermining our food, or us, on a genetic level.[3]

Nanotechnology has emerged from science fiction, science theory, and speculation to become reality in less than 20 years. When K. Eric Drexler wrote his books popularizing nanotechnology in the late 1980s, the field was still mainly speculative.[4] Today nanoscale manipulation, the arrangement of individual atoms to form very small functioning machines, is, while not yet common, becoming so. The attractive potential of nanotechnology is that a nanoscale machine—called a replicator—could build things, including other nanoscale machines. Conceivably, a million nanomachines could accomplish a task in the fraction of the time it might otherwise take: A factory could build itself and then disassemble itself when its work was done. The danger is that once these machines are released, they might not be easily stopped, and since they might consist of a few hundred molecules, they might not be easily detected. How then could we prevent runaway nanoscale replicators from turning the world into "gray goo"? In spring 2003, when Langdon Winner was asked to address a Congressional Subcommittee on the Societal Implications of Nanotechnology, he argued:

> A growing number of scientists, scholars, university administrators, and social activists express a vital interest in this topic. Clearly, there is need to initiate systematic studies of the social and ethical dimensions of nanotechnology. We need broad-ranging, detailed, intellectually rigorous inquiries conducted by persons who have no financial or institutional stake that might skew the questions raised or constrain the answers proposed.[5]

There are other challenges besides nanotechnology that need to be addressed, challenges that, while not as dire and liable to disaster as nanotechnology seems to be, pose their own unique difficulties. The emerging state of "perpetual contact"[6] generated among friends and colleagues with cell phones, text pagers, and handheld computers, suggests the possibility of emergent "smart mobs," as Howard Rheingold has described them. According to Rheingold, they promise to comprise the "next social revolution." He writes:

> Smart mobs consist of people who are able to act in concert even if they don't know each other. The people who make up smart mobs cooperate in ways never

before possible because they carry devices that possess both communication and computing capabilities.[7]

Rheingold points to a range of kinds of contact that already stakes out the shape of this revolution: Teens in places like Finland and Japan keep in contact with networks of friends throughout the day by using text messaging. Political groups organize rallies in a virtual moment's notice. Users of the "Lovegety" dating service in Japan are alerted when a Lovegety subscriber who matches their desired traits is in the immediate vicinity.

✦ James Katz calls these technologies "machines that become us," meaning machines that become integrated with our identity, and machines that are fashionable and represent social status.[8] Katz, in two volumes dedicated to these devices,[9] brings together contemporary research that explores how they transform our cultural context: Organizations change, ideas of access and privacy alter, and the feeling of lived experience is somehow different.[10] There is a dark side to this technological shift in culture. Rheingold raises three possible aspects of it: criminal exploitation of the technology, loss of privacy, and an increase in the powers of social control.[11]

How are we to understand these changes in technological culture? We must be careful not to fall back into old ways of thinking. Joy, for example, is a rather soft determinist. While he points out the social context of a technology's invention and construction, and therefore recognizes our ability to shape or limit a technology, he argues that after a certain point the situation is out of our hands and the technology becomes, in effect, autonomous. Once the Pandora's box is opened, there's no closing it.[12] Despite his conscious effort to avoid technological determinism, Rheingold still tends to describe the new technologies as arising and operating autonomously when he uses phrases such as "PCs evolved" and "then PCs mated."[13] The technologies themselves seem to arrive out of nowhere to transform social relationships. Katz likewise frames his discussion in terms of the mobile phone's and other technologies' "impacts" on our lives. Indeed, his work with Aakhus takes a turn towards expressive causality in their discussion of *Apparatgeist*, "the spirit of the machine that influences both the designs of the technology as well as the initial and subsequent significance accorded them by users, non-users, and anti-users."[14]

We have argued the limitations of these approaches throughout this book. We have also argued for a new approach with new questions. When we consider robots, genetic engineering, nanotechnology, or smart mobs, we need to ask: What are the components of the assemblage within which these technologies are developed, taken up, and used? How do these components articulate with one another? How are relations of agency configured? What is made possible? What is made invisible or impossible? What work is performed by the assemblage, in terms of identifying and discriminating on the basis of identity? How are relations of power configured? Who or what benefits? Who or what suffers? What matters? What does not? What place does space and time occupy? How does the assemblage configure affect? What shape does the feel of the world take? What ought to be changed? What ought to be supported? And, finally, how best do we proceed? What kind of technological politics would contribute to articulating the assemblage in the most desirable way? How can technological intervention contribute to making the world a better place for everyone in it?

What we are saying is that if we really want to figure out what is going on in our technological culture, and if we really want to effectively participate in and engage the decisions and processes that affect our lives, then we cannot simply respond to these new technologies with the old received approach. We cannot fall back on narratives of progress, a belief in technological determinism, a sense of autonomous technology, or retraction of the powers of creativity into the value of convenience. These old approaches do not resolve the emerging challenges of the contemporary technological culture. They will not make the world a better place in any other than the narrowest of terms for a few short-sighted people. What we need are new visions, new maps, new narratives of technology, and new articulations capable of contributing to a revitalized technological culture that we can occupy with dignity. We need, in other words, cultural studies in service of transforming technological culture.

Challenging the received view of technological culture is difficult; the old articulations are tenacious and run deep. But it is not an impossible task. Remember that this is a war of position, not a war of maneuver. We look for small changes, new maps, new stories, and rearticulations across the entire range of concepts, practices, and affects that constitute the technological assemblage. One place to begin might be to consider a term and practice that has been raised by some, a term on the basis of which some challenges have already been mounted: *limits*.

Joy concludes that "the only realistic alternative I see is relinquishment: to limit development of the technologies that are too dangerous, by limiting our pursuit of certain kinds of knowledge."[15] Winner has also pleaded for limits in the age of high technology: "I am convinced that any philosophy of technology worth its salt must eventually ask, How can we limit modern technology to match our best sense of who we are and the kind of world we would like to build?"[16] Cultural theorist Andrew Ross has asserted, "Our visions of future 'freedom' will have to be fiercely conscious of limits," and insists, along with Murray Bookchin, that "the social freedom to be derived from recognizing limits can only be attained through choice."[17]

Rearticulating our relation to limits poses a formidable task in American culture, particularly given the articulation of limits to convenience, as the pressing need to overcome, and the ultimate impossibility of overcoming, the bodily limits of space and time.[18] That limit horizon keeps on receding, and serves to taunt us to push beyond limits. What is desperately needed is a way to rearticulate limits to value something else: perhaps dignity, freedom, reciprocity, creativity, equality, and/or responsibility. Such a rearticulation would permit us to consciously agree to some limits, not in the sense of limitations that reflect our inadequacies, but limits that affirm the choice to live within something like *our means*.

We end with an example of a moment of struggle in this war of position. Both of us, Greg and Jennifer, sat on a panel at an academic conference a few years ago. The panel was titled "The Undoing Technology Roundtable." During his time to speak, Greg posed the question: Why is it so difficult to even consider getting rid of a technology? He quoted from Jerry Mander's book *Four Arguments for the Elimination of Television*: "How can I expect to succeed when even those people who loathe television find the idea of eliminating it so utterly impossible? But *why* is it so unthinkable that we might eliminate a whole technology?"[19] Jennifer then discussed the case of Japan and guns.[20] The gun was introduced to Japan by the Portuguese in 1543, and for the next hundred years the Japanese took up and perfected the technology, mass producing firearms and using more guns in battle than Europeans had at that point. Then they gave up the gun. The decision was made to stop their manufacture and use, and soon even the knowledge of how to make a gun was lost. When the Americans arrived in 1854 and forcibly opened up Japan for trade, they discovered a country almost completely without guns or knowledge of guns. The reasons for the decision to give up the gun are still under debate. Some argue that the gun was ultimately rejected because it did not fit the dignity of the samurai: Shooting posture was ungraceful, and to kill without directly facing one's opponent was dishonorable.[21] Others argue that the elite samurai class was tired of being shot by peasants: By eliminating guns, the class structure stayed intact.[22] In either case, it is an effective example of a technology that was set aside. Furthermore, throughout history innumerable technologies have, for one reason or another, fallen out of use, although it is often difficult to see what has disappeared; our cultural memory is that deficient. But most people, with just a little effort, could identify some discarded technologies such as eight-track tapes players or, possibly, nuclear power plants—after all, no new nuclear power plant has been commissioned in the United States since 1978 and it's been almost a decade since a new plant started operations.

There were various objections to "The Undoing Technology Roundtable," including an objection by a local representative from the ACLU. How, in a democratic society, could one force everyone to eliminate a technology? The only way of doing so, he maintained, would be through authoritarian means, quite at odds with our cultural commitment to democracy. In many ways he was correct; the easiest ways of getting rid of technologies trample democracy flat, because they rob all individuals of choice. But his response missed the point, which was this: Why is the question of eliminating a technology not even posed? Why is it not even among our range of choices? Shouldn't individuals have at least an option of limiting or eliminating the use of a technology? As a culture we have chosen to place limits on all sorts of things: from pollution and unfettered growth to obscene or libelous speech. Shouldn't we at least pose the question of limits to technologies? And couldn't we search for ways to do so democratically?

In this conclusion, we are not advocating the elimination of any particular technology, even if we might (we do) each have our candidates for elimination. The discussion of limits and the typical reaction to it does illustrate, however, just how unwilling people are to even consider engaging in a technological politics that might give conscious shape to technological culture. That is how powerful the received view really is. And that is also an indication of how badly we need to engage in the war of position to reconfigure the received view.

We have provided you with a cultural studies framework for understanding our technological culture and a set of concepts with which to begin to rearticulate that culture. The rest is up to you.

Notes

Introduction

1. Carey (1997), p. 316.
2. Raymond Williams (1989).

Chapter One: Progress

1. Nisbet (1980), pp. 4–5, italics removed.
2. Taylor (1947). The practice of increasing worker efficiency utilizing time and motion studies has become known as "Taylorism."
3. Noble (1982), p. xii.
4. Carey and Quirk (1989), p. 118.
5. Quoted in Marx (1987), p. 36.
6. Marx (1964).
7. Smith (1985; 1994).
8. Marx (1987).
9. Quoted in Smith (1985), p. 7.
10. Smith (1994), pp. 9–12.
11. Smith (1985).
12. Carey, "Technology and Ideology: The Case of the Telegraph," in Carey (1989), pp. 201–230.
13. Schivelbusch (1988).
14. Lyon (1994), p. 46.
15. See Moravec (1988); Dery (1996); Kurzweil (1999).
16. Negroponte (1995).
17. The discussion of the sublime draws on Nye (1994).
18. Carey and Quirk (1989).
19. Nisbet (1980).
20. Smith (1985).
21. See Piggott (2004) for documentation (including visual documentation) of the life of the Crystal Palace after the Great Exhibition.
22. Quoted in Pacey (1983), p. 25.
23. See Rogers (2003) for an insider's analysis and critique of development.
24. Pacey (1983), p. 26.
25. Noble (1982).

Chapter Two: Convenience

1. Uys (1980).
2. Tierney (1993).
3. Tierney (1993), p. 39.
4. Webster's (1976).
5. Tierney (1993), p. 40. The pre-fifteenth-century meaning of comfort as giving support or strength is still in use in the treasonous charge of "giving aid and comfort the enemy."
6. Gates (2000).
7. Cowan (1983).
8. Cowan (1983), p. 211.
9. Cowan (1983), p. 188.
10. Taylor (1947), p. 12.
11. Tierney (1993), pp. 53–57.
12. Cowan (1983), p. 101.
13. See Cowan's (1997) chapter on "Alternative Approaches to Housework," pp. 102–150.
14. Cowan (1997), p. 43.
15. Tierney (1993), p. 75.
16. Berry (1995), pp. 8, 4.
17. Rosalind Williams (2002), p. 17.

Chapter Three: Determinism

1. Izzard (1999).
2. Izzard (2000).
3. Izzard (1999).
4. Winner (1977), p. 75.
5. Lakoff and Johnson (1980), p. 69.
6. Winner (1977), p. 76.
7. Plato (360 BC/1952), p. 157.
8. Havelock (1882), pp. 6, 7.
9. Eisenstein (1979), p. 7.
10. Hughes (1994), pp. 101–113.
11. Tenner (1997), pp. 8–11.
12. Tenner (1997), p. 327.
13. Tenner (1997), p. 353.
14. Tenner (1997), p. 348.
15. The term *Hobson's choice* is said to originate with Thomas Hobson (ca. 1544–1631), of Cambridge, England, who kept a livery stable and required every customer to take either the horse nearest the stable door or none at all.
16. Pakula (1982).

Chapter Four: Control

1. Shelley (1985), although the lesson of Shelly's story is different; see below.
2. See, for example, Whale (1931; 1935); Brooks (1974); Branagh (1994); and from England's Hammer Studios: Fisher (1957; 1958); Francis (1964).
3. Winner (1977), p. 310.

4. McLuhan and Fiore (1967). This book, *The Medium Is the Massage*, is the classic statement of McLuhan's position.
5. Scheler's position is discussed in Tierney (1993), pp. 4–5.
6. See, for example, Drexler's (1986) influential predictions for the future of nanotechnology.
7. For a fascinating cultural studies perspective on this issue, see Sterne's (2003) analysis of the ways that the scientific objectification of nature and the body crystallize in the development of sound-reproduction technologies.
8. Mumford (1967), p. 191.
9. Mumford (1967), p. 192.
10. Foucault (1977), p. 205. Bentham's conception of the panopticon is considered at length by Foucault, especially pp. 195–228.
11. See, for example, Lyon's (1994) discussion of Weber's ideas.
12. Winner (1986), p. 24.
13. Tenner (1996), p. 20.
14. See Beniger (1986), especially Part II, on the crisis of control.
15. Beniger (1986), p. 227.
16. Tenner (1996).
17. Cowan (1983).
18. Tenner (1996). This and the following examples come from Tenner.
19. Karl Marx is cited here by Winner (1977), pp. 36–37.
20. Winner (1977).
21. "Industrial Society and Its Future" (1995), p. 6.
22. Joy (2000).
23. Giddens (1990).
24. Giddens (1990), p. 33.
25. Hegel (1949).
26. Winner (1977), p. 188.
27. Shelley (1985); Capek (1923).
28. Cameron (1984; 1991).
29. Wachawski and Wachawski (1999).
30. Joy (2000), p. 256.
31. William Mitchell (1995), p.   146.
32. Marvin Minsky is cited here by Riecken (1994), p. 25.

Chapter Five: Luddism

1. In addition to Thompson (1963) and Hobsbawm (1952), we draw on Thomis (1970), who has a very useful "Diary of Events, 1811–17," pp. 177–186; and the generative research that has grown out of their work: Webster and Robins (1986); Noble (1993); Sale (1995b), who also has a helpful "timeline," pp. 282–283; Robins and Webster (1999); and Fox (2002).
2. See Thompson (1963), pp. 496–497.
3. Thompson (1963), p. 543. Thomis (1970) describes the nature of this work and the machines that were targeted. A *cropper* raised the nap of finished cloth and cut it level with specialized shears (p. 15, for pictures of the shears and the process, see p. 33). The workers were replaced by the gig mill and the shearing frame (p. 50). *Stockingers*, or framework knitters, worked at frames for making hosiery and lace (p. 29, for a picture, see p. 51). *Handloom weavers* were replaced by steam looms (p. 53).
4. Thompson (1963), p. 544.
5. Thompson (1963), p. 543.

6. Both Hobsbawm (1952) and Thompson (1963) offer considerable evidence to support these multiple motives. See Thompson's chapter 15, "An Army of Redressers," pp. 472–602.
7. Hobsbawm (1952), p. 66.
8. Hobsbawm (1952), p. 58; Thompson (1963), p. 564. Some estimates are higher. Sale (1995b) suggests figures of 14,400 troops and 20,000 voluntary militia; see pp. 148–149.
9. At least 22 were hanged as Luddites (Thompson [1963], pp. 584–586). Others were killed, deported, and jailed.
10. Thompson (1963), p. 553.
11. Hobsbawm (1952), p. 59.
12. Thompson (1963), p. 549.
13. Thompson (1963), pp. 551–552.
14. Thompson (1963), p. 543.
15. Quoted in Thompson (1963), p. 554.
16. Thompson (1963), p. 552.
17. Hobsbawm (1952), p. 57.
18. See Hobsbawm (1952), pp. 66–67; Thompson (1963), pp. 601–602; and Trevelyan (1965), especially pp. 250–251, 287. Sale (1995b, p. 201) is not as generous in his assessment that "Luddism did, however, lose."
19. Sale (1995b), p. 241. His chapter on "The Neo-Luddites," pp. 237–259, does a good job of characterizing the spectrum of contemporary neo-Luddites.
20. Webster and Robins (1986).
21. Abbey (1978).
22. Fox (2002), p. 333. For the full account, see pp. 330–336, 336–364.
23. Quoted in Fox (2002), p. 335.
24. Fox (2002), p. 364.
25. Fox (2002), p. 336. For her full account of meeting those she considered modern-day Luddites, see Chapter 11, "Looking for Luddites," pp. 330–365.
26. Sale (1995b), pp. 237–238. Sale cites Glendenning (1990).

Chapter Six: Appropriate Technology

1. Many of these are taken from Dickson (1975), p. 38.
2. United Nations (1961), p. 18.
3. Daniel Lerner's book, *The Passing of Traditional Society* (1958), laid out the logic of the development mindset that devalued traditional culture in favor of the technological modern. Everett Rogers' work on *Diffusion of Innovations* extends from that tradition. See Rogers' last (2003) edition of this influential work.
4. For example, see Rybczynski (1980), pp. 10–11.
5. Shiva (1991). Also see Shiva (1989), which addresses women and development in particular.
6. Rybczynski (1980), p. 11.
7. Rybczynski (1980), p. 3.
8. Schumacher (1989).
9. Schumacher (1989), p. 186.
10. Schumacher (1989), p. 190.
11. Illich (1973).
12. This biographical information draws on Todd and La Cecla (2002); Martin (2002).
13. Illich (1973), p. 25.
14. Illich (1973), p. 22.
15. Illich (1973), p. 11.

16. Hazeltine and Bull (1999), p. 4.
17. Roszak (1978).
18. Roszak (1994).
19. Illich (1973), p. 16.
20. Rybczynski (1980), p. 159. For the original account, see Pertii (1973).
21. Rybczynski (1980), p. 160.
22. Illich (1973), p. 23.
23. Hazeltine and Bull (1999), p. 3.

Chapter Seven: The Unabomber

1. Luke (1999), p. 171.
2. This is a claim made as well by Luke (1999), p. 171.
3. Shrum (2001), p. 99.
4. See Castells (1997).
5. Corey (2000).
6. Chase (2000), p. 46.
7. See Corey (2000), pp. 180–181; Luke (1999), pp. 172–174.
8. See Corey (2000); Chase (2000).
9. Sale (1995b) discusses the neo-Luddites. Sale (1995a) considers the case of Kaczynski.
10. Strangely, despite the fact that handwritten drafts of the essay and early typescripts were found in Kaczynski's cabin, and the fact that the copy sent to the newspapers was positively typed on one of his typewriters, Kaczynski has never publicly admitted writing the essay, and neither defense nor prosecution has pressed this point (see Corey, 2000).
11. Luke (1999); Corey (2000).
12. "Industrial Society and Its Future" (1995), p. 1.
13. "Industrial Society and Its Future" (1995), p. 1.
14. Luke (1999), pp. 174–175.
15. Corey (2000), p. 159; Ellul (1964).
16. Corey (2000), p. 160.
17. Whyte (1956); Marcuse (1964).
18. Mumford (1964/1970), p. 284.
19. "Industrial Society and Its Future" (1995), p. 3.
20. Luke (1999), p. 176.
21. On this definition of technicism, see Stanley (1978), pp. xii–xiii.
22. Mumford (1964/1970), p. 291.
23. Mumford (1964/1970), p. 186.
24. Mumford (1964/1970), p. 193.
25. Chase (2000).
26. Mumford (1964/1970), especially pp. 274–276.

Chapter Eight: Defining Technology

1. *Webster's New Encyclopedic Dictionary* (2002), p. 1896.
2. Raymond Williams (1983), p. 315.
3. Romanyshyn (1989), p. 10.
4. Feenberg (1991), p. 14.
5. Winner (1986). See his discussion of "forms of life," pp. 3–18.
6. Grosz (2001), p. 182. The essay in its entirety helps the reader to develop a feel for "thingness." See pp. 166–183, 203–206.

Chapter Nine: Causality

1. Slack (1984a); Slack (1984b); Slack (1989). The four positions on causality developed in this chapter build on Slack's work.
2. Marx and Smith (1994), pp. ix–xv. The book is edited by Smith and Marx (1994).
3. Marx and Smith (1994), p. xiii.
4. Marx and Smith (1994), p. xiv.
5. Ellul (1964), p. xxv.
6. Ellul (1964), p. xxvi.
7. Ellul suggests these options in his author's foreword to the revised American edition of *The Technological Society* (1964), p. xxx.
8. Deleuze and Guattari (1987), p. 90.

Chapter Ten: Agency

1. *Webster's New World College Dictionary* (2002), pp. 25–26.
2. Latour (1993) pp. 76–82.
3. Callon and Latour (1981), p. 286.
4. Latour (1988), p. 306.
5. Latour (1999), p. 17.
6. Latour (1988), p. 303.
7. Latour (1999), p. 16.
8. Latour (1993), p. 129.

Chapter Eleven: Articulation and Assemblage

1. Murphy (2002).
2. Moyers (2002).
3. Murphy (2002).
4. Hall (1996b), p. 141.
5. Hall (1996b), p. 142.
6. Deleuze and Guattari (1987), pp. 406–407, 503–505.
7. Deleuze and Guattari (1987), p. 406.
8. Deleuze and Guattari (1987), p. 88.
9. Hall (1996b), pp. 142–143.
10. Deleuze (1995), p. 176.

Chapter Twelve: Space

1. McCarthy (2001) examines the interactions of television in public spaces. Spigel (1992) examines the interactions of television in private spaces.
2. Wise (1997), pp. xiii–xiv.
3. Lefebvre (1991), pp. 38–39.
4. Lefebvre (1991), p. 34.
5. Rheingold (2003).
6. On television, see Spigel (1992); O'Sullivan (1991); and Silverstone and Hirsch (1992). On the VCR, see Gray (1992). On computers, see Cassidy (2001) and G. Noble (1999).
7. Spigel (1992), p. 37.
8. Spigel (1992), p. 38.
9. Spigel (1992), p. 39.

10. Behl (1988), cited in Lull (1995).
11. McCarthy (2001), p. 119.
12. Picard (1997).
13. For fascinating accounts of brain-injured patients, see Damasio (1994) and Sacks (1995).
14. For an account of scaffolded thinking that explains the similarity between language and other technologies see Clark (2003), especially pp. 75-87.
15. The discussion of modes of communication, especially of the transition from orality to literacy, draws on Ong (1967; 1982); Goody (1977); and Goody and Watt (1968).
16. Innis (1950; 1964). The key section on which we draw is "The Bias of Communication" in Innis (1964), pp. 33–60.
17. Anderson (1983).
18. Headrick (2000), p. 204 discusses the development of the electronic telegraph and Morse's contribution of the code.
19. Carey and Quirk (1989).
20. See "Technology and Ideology: The Case of the Telegraph" in Carey (1989), pp. 201–230, especially pp. 217–218.
21. McLuhan and Fiore (1967).
22. Ong (1982), p. 11.

Chapter Thirteen: Identity Matters

1. Illich (1974).
2. Illich (1974), p. 119.
3. Illich (1974), p. 44.
4. See D. Noble (1986) for the importance of who owns the means of production.
5. Nader (1965; 1972).
6. Latour (1996).
7. Sclove (1995a; 1995b).
8. Sclove (1995a), pp. 91–92.
9. We draw this example from Tkach (2001); Doctors Without Borders (2001).
10. Kolko (2000), p. 218.
11. See Ritzer (1996) for a discussion of the way the system of McDonaldization works.
12. Star (1991), p. 38.
13. Star (1991), p. 39.
14. Gray (1992).
15. Gershuny (1982), cited in Gray (1992), p. 188.
16. Gray (1992), p. 248.
17. Cassidy (2001).
18. For discussion and examples of the impinging work of Moses's design work, see Caro (1974), especially p. 318. For discussion and examples of the impinging work of Moses's and others' design work see Winner's "Do Artifacts Have Politics?" in Winner (1986), pp. 19–39 and p. 180 (fn 7). For unusual evidence of Moses's explicit intentions to discriminate based on identity, see Hoving (1993), p. 245. Thomas Hoving, writing about his time as director of the New York Metropolitan Museum of Art, consulted Moses about building an underground garage. According to Hoving, Moses said: "Design it in such a way that no school buses or campers can enter. Buses drive away revenues and, besides, all bus drivers pocket the money they got for parking." Campers had to be discouraged because "squatters will stay for life." Moses's solution, which was adopted by the museum, was to lower the height of the entrance to the garage.
19. Kolko (2000).

20. Both Hill (2001) and Nakamura (2002) argue that minorities need to create media and media texts.
21. See the 2001 Academy Award nominee for Best Documentary feature, *The Sound and the Fury*, directed by Aronson (2001).
22. See Bowker and Star (1999), especially the introduction, pp. 1–32.
23. Nussbaum (2000).
24. For an example of work on the social construction of race, see Omi and Winant (1986).
25. See, for example, Bowker and Star (1999) on race classification under apartheid, pp. 195–225. See also Omi and Winant (1993), in which they argue that the fact that race is socially constructed does not mean that it is pure ideology. Rather, race is "a fundamental principle of social organization and identity formation" that is always relational and in process, pp. 5–6.
26. Balsamo (1996).
27. Yamamoto (1999).
28. See "Technologies as Forms of Life" in Winner (1986), pp. 3–18.

Chapter Fourteen: Challenging Identity

1. Balsamo (1996), pp. 41–55.
2. Turkle (1995).
3. Turkle (1995), p. 13. The first and third editorial brackets are by the authors; the second is by Turkle.
4. Spender (1995), p. 244.
5. Marvin (1988), pp. 69–70.
6. *The World Almanac and Book of Facts* (2003), p. 708.
7. Spender (1995), p. 244.
8. The discussion of race in cyberspace draws on Kolko, Nakamura, and Rodman (2000).
9. King (1963).
10. *Webster's Dictionary of the English Language* (1987), p. 1130.
11. This is Greek philosopher Protagoras's most famous phrase, spoken 2,500 years ago and discussed by Plato in the *Theaetetus*, See Burnyeat (1990).
12. The riddle goes something like this: A father rushes his child to the emergency room, where the Dr. Smith exclaims, "I can't operate—it's my son!" How can this be? We are baffled until we can break through the cultural default and understand that the doctor can be a woman. (There is, of course, a whole other set of cultural defaults this riddle raises with respect to stepfamilies.)
13. Kolko, Nakamura, and Rodman (2000), pp. 4–5.
14. Nakamura (2002).
15. Haraway (1985).
16. Haraway (1985), p. 66, emphasis in the original.
17. Lucas (2001).
18. For a discussion of the concept that we have always been cyborg, see Clark (2003).
19. Turing (1950).
20. Platt (1995), p. 180.
21. See Turkle (1995) and Leonard (1997).
22. Turkle (1995).
23. Boden (1996), p. 1.
24. Langton (1996), p. 40.
25. Balsamo (1996), pp. 80–115; Cartwright (1998); and Mitchell and Georges (1998) raise this and related issues.

Chapter Fifteen: Politics

1. Winner (1986), p. 22.
2. Sclove (1995a), p. 89.
3. Winner (1986), p. 20.
4. Winner (1986), pp. 28–29.
5. Winner (1986), p. 10.
6. The following discussion draws on the ideas Winner develops in *The Whale and the Reactor* (1986). Where something specific is referred to, we cite specific pages.
7. See the section on "Technology: Reform and Revolution," in Winner (1986), pp. 61–117.
8. Slack (1984c), especially pp. 251–252.
9. Winner (1986), p. 137. The entire essay, "The State of Nature Revisited" (pp. 121–137), is relevant here.
10. See Slack (1998) for an example of the range of political uses to which an appeal to nature can be put.
11. The variable use of nature in cultural politics is addressed by Evernden (1992); Slack and Whitt (1992); and Whitt and Slack (1994).
12. Winner (1986), p. 47.
13. Winner (1986), p. 47.
14. Winner (1986), p. 47.
15. Winner (1986), p. 48.
16. Winner (1986), p. 48.
17. Winner (1986), p. 48.
18. Winner (2003).
19. Barber (1984).
20. Sclove (1995a; 1995b).
21. Sclove (1995b), p. 98.
22. Slack (1996).
23. Sclove (1995b), p. 156.
24. Pacey (1983), p. 26.
25. Gramsci (1971), pp. 229–239. Hall (1996a) discusses the war of maneuver and the war of position in relation to cultural studies; see especially pp. 426–428.
26. Balsamo (1996), p. 161.
27. On justice and dignity, see Winner (1986); on Buddhist principles, see Pacy (1983) and Schumacher (1989); on principles of strong democracy, see Sclove (1995a; 1995b).

Chapter Sixteen: Globalization

1. The descriptions of Vishwakarma Puja and Ratha-yâtrâ are based on Greg's personal knowledge synthesized with these sources: *Oxford English Dictionary* (online); *The New Encyclopedia Britannica* (1994) on Jagannâth; Bangalinet (2003) and About, Inc. (2004) on Vishwakarma Puja. All accessed January 2004.
2. *Oxford English Dictionary* (online).
3. Sundaram (2000).
4. Sundaram (2001).
5. du Gay (2000), p. 115.
6. For a description of the SITE project, see Chander and Karnik (1976).
7. Appadurai (1996).
8. Appadurai (1996), p. 33
9. Mies and Shiva (1993).
10. Shiva (1995), pp. 211–212.

11. Starr (2000). The quotation is from p. 67.
12. Meikle (2002), pp. 7–8.

Conclusion

1. Joy (2000).
2. Lovins and Lovins (2000).
3. Virlio (2000).
4. Drexler (1986); Drexler, Peterson, and Pergamit (1991). Nanotechnology was actually a theory derived from Richard Feynman (1959) in a speech titled "There's Plenty of Room at the Bottom" at the annual meeting of the American Physical Society.
5. Winner (2003).
6. Katz and Aakhus (2002).
7. Rheingold (2003), p. xii.
8. Katz (2003).
9. Katz (2003); Katz and Aakhus (2002).
10. Katz and Aakhus (2002), p. xxi.
11. Rheingold (2003), see pp. xxi and xxii and his Chapter 8.
12. Joy (2000), p. 256.
13. Rheingold (2003) uses these kinds of phrases throughout his book. See Rheingold (2000), p. 376, for his intentions regarding technological determinism.
14. Katz and Aakhus (2002), p. 305.
15. Joy (2000), p. 254.
16. Winner (1986), p. xi.
17. Ross (1991), pp. 169, 5.
18. We note, however, that in all seriousness, scientists engaged in genetic engineering are considering the possibility that it might be possible to "make an organism immortal." See, for example, the interview with biologist Cynthia Kenyon (2004) about her research that has extended the lifespan of worms.
19. Mander (1978), p. 348.
20. She told the story as told by Perrin (1979).
21. Perrin (1979), p. 45.
22. Sclove (1995), p. 56.

References

Abbey, Edward. 1978. *The monkey wrench gang*. Edinburgh, UK: Canongate Publishing.

About, Inc. 2004. Vishwakarma and mythical buildings: The divine craftsman and his architectural marvels. Hinduism.about.com/library/weekly/aa092401a.htm. Accessed January 2004.

Anderson, Benedict. 1983. *Imagined communities: Reflections on the origin and spread of nationalism*. New York, NY: Verso.

Appadurai, Arjun. 1996. *Modernity at large: Cultural dimensions of globalization*. Minneapolis, MN: University of Minnesota Press.

Aronson, Josh, director. 2001. *The sound and the fury*, produced by Roger Weisberg. PBS. Available at www.pbs.org/wnet/soundandfury/index.html. Accessed November 24, 2004.

Balsamo, Anne. 1996. *Technologies of the gendered body: Reading cyborg women*. Durham, NC: Duke University Press.

Bangalinet. 2003. Vishwakarma uja. www.bangalinet.com/vishwakarmapuja.htm. Accessed January 2004.

Barber, Benjamin. 1984. *Strong democracy: Participatory politics for a new age*. Berkeley, CA: University of California Press.

Behl, Neena. 1988. Equalizing status: Television and tradition in an Indian village. In *World families watch television*, edited by James Lull. Newbury Park, CA: Sage Publications, pp. 136-157.

Beniger, James R. 1986. *The control revolution: Technological and economic origins of the information society*. Cambridge, MA: Harvard University Press.

Berry, Wendell. 1995. *Another turn of the crank*. Washington, DC: Counterpoint.

Boden, Margaret A. 1996. Introduction. In *The philosophy of artificial life*, edited by Margaret Boden. New York, NY: Oxford University Press, pp. 1-35.

Bowker, Geoffrey C. and Susan Leigh Star. 1999. *Sorting things out: Classification and its consequences*. Cambridge, MA: MIT Press.

Branagh, Kenneth, director. 1994. *Mary Shelley's Frankenstein*. American Zoetrope.

Brooks, Mel, director. 1974. *Young Frankenstein*. 20th Century Fox.

Burnyeat, Myles. 1990. *The Theaetetus of Plato with a translation of Plato's Theaetetus by M. J. Levett*, revised by Myles Burnyeat. Indianapolis, IN: Hackett.

Callon, Michel and Bruno Latour. 1981. Unscrewing the big Leviathan: How actors macro-structure reality and how sociologists help them do so. In *Advances in social theory and methodology: Toward an integration of micro- and macro-sociology*, edited by K. Knorr-Cetina and A.V. Cicourel. Boston, London and Henley: Routledge and Kegan Paul, pp. 277-303.

Cameron, James, director. 1984. *The terminator*. Hemdale Film Corporation.

Cameron, James, director. 1991. *Terminator 2: Judgment day*. Carolco Pictures, Inc.

Capek, K. 1923. *R.U.R. (Rossum's universal robots): A fantastic melodrama*, translated by P. Selver. Garden City, NY: Doubleday.

Carey, James W. 1989. *Communication as culture: Essays in media and society*. Boston, MA: Unwin Hyman.

Carey, James W. 1997. Afterword: The culture in question. In *James Carey*, edited by Eve Stryker Munson and Catherine A. Warren. Minneapolis, MN: University of Minnesota Press, pp. 308-339.

Carey, James W. and John J Quirk. 1989. The mythos of the electronic revolution. In *Communication as culture: Essays on media and society*, by James W. Carey. Boston, MA: Unwin Hyman, pp. 113-141.

Caro, Robert A. 1974. *The power broker: Robert Moses and the fall of New York*. New York, NY: Knopf.

Cartwright, Elizabeth. 1998. The logic of heartbeats: Electronic fetal monitoring and biomedically constructed birth. In *Cyborg babies: From techno-sex to techno-tots*, edited by Robbie Davis-Floyd and Joseph Dumit. New York, NY: Routledge, pp. 240-254.

Cassidy, Marsha. 2001. Cyberspace meets domestic space: Personal computers, women's work, and the gendered territories of the family home. *Critical Studies in Media Communication*, 18(1): 44-65.

Castells, Manuel. 1997. *The information age: Economy, society and culture: Volume II: The power of identity*. Malden, MA: Blackwell.

Chander, Romesh and Kiran Karnik. 1976. *Planning for satellite broadcasting: The Indian instructional televison experiment*. Paris, France: UNESCO.

Chase, Alston. 2000. Harvard and the making of the Unabomber. *The Atlantic Monthly*, June, pp. 41-65.

Clark, Andy. 2003. *Natural-born cyborgs: Minds, technologies, and the future of human intelligence*. New York, NY: Oxford University Press.

Corey, Scott. 2000. On the Unabomber. *Telos*, 118: 157-182.

Cowan, Ruth Schwartz. 1983. *More work for mother: The ironies of household technology from the open hearth to the microwave*. New York, NY: Basic Books.

Cowan, Ruth Schwartz. 1997. *A social history of American technology*. New York, NY: Oxford University Press.

Damasio, Antonio R. 1994. *Descartes' error: Emotion, reason and the human brain.* New York, NY: Putnam.

Deleuze, Gilles. 1995. *Negotiations: 1972-1990,* translated by Martin Joughin. New York, NY: Columbia University Press.

Deleuze, Gilles and Félix Guattari. 1987. *A thousand plateaus: Capitalism and schizophrenia,* translated by Brian Massumi. Minneapolis, MN: University of Minnesota Press. (Orig. 1980.)

Dery, Mark. 1996. *Escape velocity: Cyberculture at the end of the century.* New York, NY: Grove Press.

Dickson, David. 1975. *The politics of alternative technology.* New York, NY: Universe Books.

Doctors Without Borders/Médecins Sans Frontiéres (MSF). 2001. Supply of sleeping sickness drugs secured. http://www.doctorswithoutborders.org/pr/2001/05-03-2001.shtml. Accessed January 31, 2004.

Drexler, K. Eric. 1986. *Engines of creation: The coming era of nanotechnology.* New York, NY: Anchor Books.

Drexler, K. Eric, Chris Peterson with Gayle Pergamit. 1991. *Unbounding the future: The nanotechnology revolution: The path to molecular manufacturing and how it will change our world.* New York, NY: Quill.

du Gay, Paul. 2000. Representing "globalization": Notes on the discursive orderings of economic life. In *Without guarantees: In honour of Stuart Hall,* edited by Paul Gilroy, Lawrence Grossberg, and Angela McRobbie. New York, NY: Verso, pp. 113-125.

Eisenstein, Elizabeth. 1979. *The printing press as an agent of change: Communications and cultural transformations in early-modern Europe.* Cambridge, UK: Cambridge University Press.

Ellul, Jacques. 1964. *The technological society,* translated by John Wilkinson. New York, NY: Vintage.

Evernden, Neil. 1992. *The social creation of nature.* Baltimore, MD: Johns Hopkins University Press.

Feenberg, Andrew. 1991. *Critical theory of technology.* New York, NY: Oxford University Press.

Feynman, Richard. 1959. There's plenty of room at the bottom. Talk given on December 29 at the annual meeting of the American Physical Society at the California Institute of Technology. Published in *Engineering and Science,* 23 (1960): 22-36. Transcribed at http://www.zyvex.com/nanotech/feynman.html.

Fisher, Terence, director. 1957. *The curse of Frankenstein.* Hammer Film Productions Limited.

Fisher, Terence, director. 1958. *The revenge of Frankenstein.* Hammer Film Productions Limited.

Foucault, Michel. 1977. *Discipline and punish: The birth of the prison,* translated by Alan Sheridan. Harmondsworth, UK: Peregrine Books.

Fox, Nicols. 2002. *Against the machine: The hidden Luddite tradition in literature, art, and individual lives*. Washington, DC: Island Press.

Francis, Freddie, director. 1964. *The evil of Frankenstein*. Hammer Film Productions Limited.

Gates, Bill. 2000. *Business at the speed of thought: Succeeding in the digital economy*. New York, NY: Warner Books.

Gershuny, J. L. 1982. Household tasks and the use of time. In *Living in South London*, edited by S. Wallman and Associates. London: Gower.

Giddens, Anthony. 1990. *The consequences of modernity*. Stanford, CA: Stanford University Press.

Gleick, James. 2000. *Faster: The acceleration of just about everything*. New York, NY: Vintage.

Glendenning, Chellis. 1990. Notes toward a neo-Luddite manifesto. *Utne Reader*, March, pp. 50-53.

Goody, Jack. 1977. *The domestication of the savage mind*. Cambridge, UK: Cambridge University Press.

Goody, Jack and Ian Watt. 1968. The consequences of literacy. In *Literacy in traditional societies*, edited by Jack Goody. Cambridge, UK: Cambridge University Press, pp. 27-68.

Gramsci, Antonio. 1971. *Selections for the prison notebooks*, edited and translated by Quintin Hoare and Geoffrey Nowell Smith. New York, NY: International Publishers.

Gray, Ann. 1992. *Video playtime: The gendering of a leisure technology*. New York, NY: Routledge.

Grosz, Elizabeth. 2001. *Architecture from the outside: Essays on virtual and real space*. Cambridge, MA: MIT Press.

Hall, Stuart. 1996a. Gramsci's relevance for the study of race and ethnicity. In *Stuart Hall: Critical dialogues in cultural studies*, edited by David Morley and Kuan-Hsing Chen. New York, NY: Routledge, pp. 411-440.

Hall, Stuart. 1996b. On postmodernism and articulation: An interview with Stuart Hall, edited by Lawrence Grossberg. In *Stuart Hall: Critical dialogues in cultural studies*, edited by David Morley and Kuan-Hsing Chen. New York, NY: Routledge, pp. 131-150.

Haraway, Donna. 1985. A manifesto for cyborgs: Science, technology, and socialist feminism in the 1980s. *Socialist Review*, 80: 65-107.

Havelock, Eric A. 1982. *The literate revolution in Greece and its cultural consequences*. Princeton, NJ: Princeton University Press.

Hazeltine, Barrett and Christopher Bull. 1999. *Appropriate technology: Tools, choices, and implications*. San Diego, CA: Academic Press.

Headrick, Daniel. 2000. *When information came of age: Technologies of knowledge in the age of reason and revolution, 1700-1850*. New York, NY: Oxford University Press.

Hegel, Georg Wilhelm Friedrich. 1949. *The phenomenology of mind*. 2nd edition, translated by J. B. Baillie. London: G. Allen & Unwin. (Orig. 1807)

Hill, Logan. 2001. Beyond access: Race, technology, community. In *Technicolor: Race, technology and everyday life*, edited by Alondra Nelson, Thuy Linh N. Tu with Alicia Headlam Hines. New York, NY: New York University Press, pp. 13-33.

Hughes, Thomas P. 1994. Technological momentum. In *Does technology drive history? The dilemma of technological determinism*, edited by Merrit Roe Smith and Leo Marx. Cambridge, MA: MIT Press, pp.101-113.

Hobsbawm, E. J. 1952. The machine breakers. *Past and present*, 1: 57-70.

Hoving, Thomas. 1993. *Making the mummies dance: Inside the Metropolitan Museum of Art*. New York, NY: Simon & Schuster.

Illich, Ivan. 1973. *Tools for conviviality*. New York, NY: Harper & Row.

Illich, Ivan. 1974. *Energy and equity*. Boston, MA: Marion Boyars.

Industrial society and its future. 1995. Supplement to the *Washington Post*, September 19. Attributed to Theodore Kaczynski.

Innis, Harold A. 1950. *Empire and communications*. Toronto: University of Toronto Press.

Innis, Harold A. 1964. *The bias of communication*. Toronto: University of Toronto Press. (Orig. 1951.)

Izzard, Eddie. 1999. *Dress to kill*. Ella Publications. [DVD].

Izzard, Eddie. 2000. *Circle*. Henry Fonda Theater, Los Angeles. June 17. [Performance].

Joy, Bill. 2000. Why the future doesn't need us. *Wired*, 8 (4) 238-262.

Katz, James, ed. 2003. *Machines that become us: The social context of personal communication technology*. London: Transaction Publishers.

Katz, James and Mark Aakhus, eds. 2002. *Perpetual contact: Mobile communication, private talk, public performance*. New York, NY: Cambridge University Press.

Kenyon, Cynthia. 2004. The biologist who extends life spans. Interviewed by David Ewing Duncan. *Discover*, 25 (3), pp. 16, 18-19.

King, Martin Luther, Jr. 1963. I have a dream. Address delivered at the march on Washington for jobs and freedom. Washington, DC, August 28.

Kurzweil, Ray. 1999. *The age of spiritual machines*. New York, NY: Penguin.

Kolko, Beth E. 2000. Erasing @race: Going white in the (inter)face. In *Race in cyberspace*, edited by Beth E. Kolko, Lisa Nakamura, and Gilbert Rodman. New York, NY: Routledge, pp. 213-232.

Kolko, Beth E., Lisa Nakamura, and Gilbert Rodman. 2000. Race in cyberspace: An introduction. In *Race in cyberspace*, edited by Beth E. Kolko, Lisa Nakamura, and Gilbert Rodman. New York, NY: Routledge, pp. 1-13.

Lakoff, George and Mark Johnson. 1980. *Metaphors we live by*. Chicago, IL: University of Chicago Press.

Langton, Christopher G. 1996. Artificial life. In *The philosophy of artificial life*, edited by Margaret Boden. New York, NY: Oxford University Press, pp. 39-94.

Latour, Bruno [as Jim Johnson]. 1988. Mixing humans and nonhumans together: The sociology of a door-closer. *Social Problems*, 35(3): 298-310.

Latour, Bruno. 1993. *We have never been modern*, translated by Catherine Porter. Cambridge, MA: Harvard University Press.

Latour, Bruno. 1996. *Aramis, or the love of technology*, translated by Catherine Porter. Cambridge, MA: Harvard University Press.

Latour, Bruno. 1999. On recalling ANT. In *Actor network theory and after*, edited by John Law and John Hassard. Malden, MA: Blackwell, pp. 15-25.

Lefebvre, Henri. 1991. *The production of space*, translated by Donald Nicholson-Smith. Cambridge, MA: Blackwell.

Leonard, Andrew. 1997. *Bots: The origin of new species*. San Francisco, CA: Hardwired.

Lerner, Daniel. 1958. *The passing of traditional society: Modernizing the Middle East*. Glencoe, IL: Free Press.

Lovins, Amory B. and L. Hunter Lovins. 2000. A tale of two botanies. *Wired*, 8 (4), 247.

Lucas, Thomas, director. 2001. Beyond human. Alexandria, VA: PBS Home Video.

Luke, Timothy. 1999. Slowburn, fast detonation, killer fragments: Rereading the Unabomber manifesto. In *Capitalism, democracy, and ecology: Departing from Marx*. Urbana, IL: University of Illinois Press, pp. 171-195.

Lull, James. 1995. *Media, communication, culture: A global approach*. New York, NY: Columbia University Press.

Lyon, David. 1994. *The electronic eye: The rise of surveillance society*. Minneapolis, MN: University of Minnesota Press.

Mander, Jerry. 1978. *Four arguments for the elimination of television*. New York, NY: Quill.

Marcuse, Herbert. 1964. *One-dimensional man*. Boston, MA: Beacon Press.

Martin, Douglas. 2002. Ivan Illich, 76, philosopher who challenged status quo, is dead. *New York Times*, December 4. http://www.NYTimes.com. Accessed February 10, 2003.

Marvin, Carolyn. 1988. *When old technologies were new: Thinking about electric communication in the late nineteenth century*. New York, NY: Oxford University Press.

Marx, Leo. 1964. *The machine in the garden: Technology and the pastoral ideal in America*. New York, NY: Oxford University Press.

Marx, Leo. 1987. Does improved technology mean progress? *Technology Review*, January, pp. 32-41, 71.

Marx, Leo and Merritt Roe Smith, eds. 1994. Introduction. In *Does technology drive history? The dilemma of technological determinism*, edited by Merrit Roe Smith and Leo Marx. Cambridge, MA: MIT Press, pp. ix-xv.

McCarthy, Anna. 2001. *Ambient television: Visual culture and public space*. Durham, NC: Duke University Press.

McLuhan, Marshall and Quentin Fiore. 1967. *The medium is the massage: An inventory of effects*. New York, NY: Bantam Books.

Meikle, Graham. 2002. *Future active: Media activism and the internet*. New York, NY: Routledge.

Mies, Maria and Vandana Shiva. 1993. *Ecofeminism*. London: Zed Books.

Mitchell, Lisa M. and Eugenia Georges. 1998. Baby's first picture: The cyborg fetus of ultrasound imaging. In *Cyborg babies: From techno-sex to techno-tots*, edited by Robbie Davis-Floyd and Joseph Dumit. New York, NY: Routledge, pp. 105-124.

Mitchell, William. 1995. *City of bits*. Cambridge, MA: MIT Press.

Moravec, Hans. 1988. *Mind children: The future of robot and human intelligence*. Cambridge, MA: Harvard University Press.

Moyers, Bill. 2002. *NOW with Bill Moyers*. PBS. Airdate: September 13. [TV Program].

Mumford, Lewis. 1964/1970. *The myth of the machine: The pentagon of power*. New York, NY: Harcourt, Brace, Jovanovich.

Mumford, Lewis. 1967. *The myth of the machine: Technics and human development: Volume one*. New York, NY: Harcourt, Brace, Jovanovich.

Murphy, Dean E. 2002. As security cameras sprout, someone's always watching. *New York Times*, September 29. [online]. Accessed September 29, 2002.

Nader, Ralph. 1965. *Unsafe at any speed: The designed-in dangers of the American automobile*. New York, NY: Grossman.

Nader, Ralph. 1972. *Unsafe at any speed: The designed-in dangers of the American automobile (Expanded edition)*. New York, NY: Grossman.

Nakamura, Lisa. 2002. *Cybertypes: Race, ethnicity, and identity on the internet*. New York, NY: Routledge.

Negroponte, Nicholas. 1995. *Being digital*. New York, NY: Knopf.

New Encyclopedia Britannica, Volume 6. (1994). Jagannatha. Chicago: Encyclopedia Britannica, p. 469.

Nisbet, Robert. 1980. *History of the idea of progress*. New York, NY: Basic Books.

Noble, David. 1982. Introduction. In *Architect or bee? The human/technology relationship*, by Mike Cooley, edited by Shirley Cooley. Boston, MA: South End Press, pp. xi-xxi.

Noble, David. 1986. *Forces of production: A social history of industrial automation*. New York, NY: Oxford University Press.

Noble, David. 1993. *Progress without people: In defense of Luddism*. Chicago, IL: Charles H. Kerr Publishing.

Noble, Greg. 1999. Domesticating technology: Learning to live with your computer. *Australian Journal of Communication*, 26(2), 59-76.

Nussbaum, Emily. 2000. A question of gender. *Discover*, January, pp. 92-99.

Nye, David E. 1994. *American technological sublime*. Cambridge, MA: MIT Press.

O'Sullivan, Tim. 1991. Television memories and cultures of viewing, 1950-65. In *Popular television in Britain: Studies in cultural history*, edited by John Corner. London: British Film Institute, pp. 159-181.

Omi, Michael and Howard Winant. 1986. *Racial formation in the United States: From the 1960s to the 1980s*. New York, NY: Routledge.

Omi, Michael and Howard Winant. 1993. On the theoretical status of the concept of race. In *Race, identity, and education in education*, edited by Cameron McCarthy and Warren Crichlow. New York: NY: Routledge, pp. 3-10.

Ong, Walter. 1967. *The Presence of the word: Some prolegomena for cultural and religious history*. Minneapolis, MN: University of Minnesota Press.

Ong, Walter. 1982. *Orality and literacy: The technologizing of the word*. New York, NY: Routledge.

Oxford English Dictionary [online]. www.OED.com. Accessed June 6, 2003.

Pacey, Arnold. 1983. *The culture of technology*. Cambridge, MA: MIT Press.

Pakula, Alan J., director. 1982. *Sophie's Choice*. Incorporated Television Company.

Pelto, Pertii J. 1973. *The snowmobile revolution: Technology and social change in the Arctic*. Menlo Park, CA: Cummings.

Perrin, Noel. 1979. *Giving up the gun: Japan's reversion to the sword, 1543-1879*. Boston, MA: David R. Godine.

Picard, Rosalind W. 1997. *Affective computing*. Cambridge, MA: MIT Press.

Piggott, J. R. 2004. *Palace of the people: The crystal palace at Sydenham: 1854-1936*. Madison, WI: University of Wisconsin Press.

Plato. 1952. *Phaedrus*, translated by R. Hackforth. Indianapolis, IN: Bobbs-Merrill.

Platt, Charles. 1998. What's it mean to be human, anyway? In *Composing cyberspace: Identity, community, and knowledge in the electronic age*, edited by Richard Holeton. New York, NY: McGraw-Hill, pp. 12-20.

Rheingold, Howard. 2000. *The virtual community: Homesteading on the electronic frontier*, revised edition. Cambridge, MA: MIT Press.

Rheingold, Howard. 2003. *Smart mobs: The next social revolution*. New York, NY: Perseus.

Riecken, D. 1994. A conversation with Marvin Minsky about agents. *Communications of the ACM*, 37 (7), pp. 23-9.

Ritzer, George. 1996. *The McDonaldization of society: An investigation into the changing character of contemporary social life*. Thousand Oaks, CA: Pine Forge Press.

Robins, Kevin and Frank Webster. 1999. *Times of the technoculture: From the information society to the virtual life*. London: Routledge.

Rogers, Everett M. 2003. *Diffusion of Innovations*, 5th edition. New York, NY: Free Press.

Romanyshyn, Robert D. 1989. *Technology as symptom and dream*. New York, NY: Routledge.

Ross, Andrew. 1991. *Strange weather: Culture, science and technology in the age of limits*. New York, NY: Verso.

Roszak, Theodore. 1978. *Person/planet: The creative disintegration of industrial society*. Garden City, NY: Doubleday.

Roszak, Theodore. 1994. *The cult of information: A neo-Luddite treatise on high tech, artificial intelligence, and the true art of thinking*, 2nd edition. Berkeley, CA: University of California Press.

Rybczynski, Witold. 1980. *Paper heroes: A review of appropriate technology*. Garden City, NY: Anchor Books.

Sacks, Oliver. 1995. *An anthropologist on Mars: Seven paradoxical tales*. New York, NY: Alfred A. Knopf.

Sale, Kirkpatrick. 1995a. Is there a method to his madness? *The Nation*, September 25, pp. 305-311.

Sale, Kirkpatrick. 1995b. *Rebels against the future: The Luddites and their war on the industrial revolution: Lessons for the computer age*. New York, NY: Addison-Wesley.

Schivelbusch, Wolfgang. 1988. *Disenchanted night: The industrialization of light in the nineteenth century*, translated by Angela Davies. Berkeley, CA: University of California Press.

Schumacher, Ernst Friedrick. 1989. *Small is beautiful: Economics as if people mattered*. New York, NY: Harper & Row. (Orig. 1973)

Sclove, Richard. 1995a. Making technology democratic. In *Resisting the virtual life: The culture and politics of information*, edited by James Brook and Iain Boal. San Francisco, CA: City Lights, pp. 85-101.

Sclove, Richard. 1995b. *Democracy and technology*. New York, NY: Guilford.

Shelley, Mary Wollstonecraft. 1985. *Frankenstein, or the modern Prometheus*. New York, NY: Penguin Books. (Orig. 1818)

Shiva, Vandana. 1989. *Staying alive: Women, ecology and development*. London: Zed Books.

Shiva, Vandana. 2001. The green revolution in the Punjab. *The Ecologist*, 21 (2). Accessed at http://livingheritage.org/green-revolution.htm. Accessed February 8, 2004.

Shiva, Vandana. 1995. Biotechnological development and the conservation of biodiversity. In *Biopolitics: A feminist and ecological reader on biotechnology*, edited by Vandana Shiva and Ingunn Moser. London, UK: Zed Books, pp. 93-213.

Shrum, Wesley. 2000. We were the Unabomber. *Science, Technology, and Human Values*, 26 (1), 90-101.

Silverstone, Roger and Hirsch, Eric, eds. 1992. *Consuming technologies: Media and information in domestic spaces*. New York, NY: Routledge.

Slack, Jennifer Daryl. 1984a. *Communication technologies and society: Conceptions of causality and the politics of technological intervention*. Norwood, NJ: Ablex.

Slack, Jennifer Daryl. 1984b. Surveying the impacts of communication technologies. In *Progress in communication sciences: Volume 5*, edited by Brenda Dervin and Melvin J. Voigt. Norwood, NJ: Ablex, pp. 73-109.

Slack, Jennifer Daryl. 1984c. The information revolution as ideology. *Media, culture and society*, 6, pp. 247-256.

Slack, Jennifer Daryl. 1989. Contextualizing technology. In *Rethinking communication: Volume 2: Paradigm exemplars*, edited by Brenda Dervin, Lawrence Grossberg, Barbara J. O'Keefe, and Ellen Wartella. Newbury Park, CA: Sage, pp. 329-345.

Slack, Jennifer Daryl. 1996. The theory and method of articulation in cultural studies. In *Stuart Hall: Critical dialogues in cultural studies*, edited by David Morley and Kuan-Hsing Chen. New York, NY: Routledge, pp. 112-127.

Slack, Jennifer Daryl. 1998. The politics of the pristine. *Topia: Canadian Journal of Cultural Studies*, 2: 67-90.

Slack, Jennifer Daryl and Laurie Anne Whitt. 1992. Ethics and cultural studies. In *Cultural studies*, edited by Lawrence Grossberg, Cary Nelson, and Paula Treichler. New York, NY: Routledge, pp. 571-592.

Smith, Merritt Roe. 1985. Technology, industrialization, and the idea of progress in America. In *Responsible science: The impact of technology on society*, edited by Kevin B. Byrne. San Francisco, CA: Harper & Row, pp. 1-30.

Smith, Merritt Roe. 1994. Technological determinism in American culture. In *Does technology drive history? The dilemma of technological determinism*, edited by Merritt Roe Smith and Leo Marx. Cambridge, MA: MIT Press, pp. 1-35.

Smith, Merritt Roe and Leo Marx, eds. 1994. *Does technology drive history? The dilemma of technological determinism*. Cambridge, MA: MIT Press.

Spender, Dale. 1995. *Nattering on the Net: Women, power and cyberspace*. North Melbourne, Australia: Spinifex.

Spigel, Lynn. 1992. *Make room for TV: Television and the family ideal in postwar America*. Chicago, IL: University of Chicago Press.

Stanley, Manfred. 1978. *The technological conscience: Survival and dignity in an age of expertise*. Chicago, IL: University of Chicago Press.

Star, Susan Leigh. 1991. Power, technology, and the phenomenology of conventions: On being allergic to onions. In *A sociology of monsters? Power, technology, and the modern world*, edited by John Law. Oxford: Blackwell, pp. 27-55.

Starr, Amory. 2000. *Naming the enemy: Anti-corporate movements confront globalization*. London: Zed Books

Sterne, Jonathan. 2003. *The audible past: Cultural origins of sound reproduction*. Durham, NC: Duke University Press.

Sundaram, Ravi. 2000. Beyond the nationalist panopticon: The experience of cyberpublics in India. In *Electronic media and technoculture*, edited by J.T. Caldwell. New Brunswick, NJ: Rutgers University Press, pp. 270-294.

Sundaram, Ravi. 2001. Recycling modernity: Pirate electronic cultures in India. In *Sarai Reader 01: The public domain*. New Delhi: Sarai, pp. 93-99.

Taylor, Frederick W. 1947. *The principles of scientific management*. New York, NY: W. W. Norton. (Orig. 1911)

Tenner, Edward. 1996. *Why things bite back: Technology and the revenge of unintended consequences*. New York, NY: Vintage.

Thomis, Malcolm I. 1970. The Luddites: Machine-breaking in regency England. Newton Abbot, UK: David & Charles Publishers.

Thompson, E. P. 1963. *The making of the English working class*. New York, NY: Vintage.

Tierney, Thomas F. 1993. *The value of convenience: A genealogy of technical culture*. Albany, NY: State University of New York Press.

Tkach, Andrew, producer. 2001. Sleeping sickness. *60 minutes*. CBS. Airdate: February 11 [TV Program].

Todd, Andrew and Franco La Cecla. 2002. A polymath and polemicist, his greatest contribution was as an archaeologist of ideas, rather than an ideologue. *The Observer*, December 8. Accessed via Guardian Unlimited: http:/www.guardian. co.uk. Accessed February 6, 2003.

Trevelyan, G. M. 1965. *British history in the nineteenth century and after: 1782-1919*. Harmondsworth, UK: Penguin.

Turing, A. M. 1950. Computing machinery and intelligence. *Mind*, 49: 433-460.

Turkle, Sherry. 1995. *Life on the screen: Identity in the age of the internet*. New York, NY: Simon & Schuster.

United Nations. 1961. 1710 (XVI). United Nations development decade: A programme for international economic co-operation. Accessed at http://ods-dds-ny.un.org/doc/RESOLUTION/GEN/NRO/167/63/IMG/NR016763. pdf?OpenElement. Accessed February 8, 2004.

Uys, Jamie, director. 1980. *The gods must be crazy*. CAT Films.

Virilio, Paul. 2000. *The information bomb*, translated by Chris Turner. New York, NY: Verso.

Wachowski, Andy and Larry Wachowski, directors. 1999. *The matrix*. Groucho II Film Partnership.

Webster, Frank and Kevin Robins. 1986. *Information technology: A Luddite analysis*. Norwood, NJ: Ablex.

Webster's dictionary of the English language, new lexicon edition. 1987. New York, NY: New Lexicon Publications.

Webster's new encyclopedic dictionary. 2002. Springfield, MA: Merriam-Webster.

Webster's new world college dictionary, 4th edition. 2002. Cleveland, OH: Wiley.

Webster's ninth new collegiate dictionary. 1990. Cleveland, OH: Wiley.

Webster's third new international dictionary. 1976. Cleveland, OH: Wiley.

Whale, James, director. 1931. *Frankenstein*. Universal Pictures.

Whale, James, director. 1935. *Bride of Frankenstein*. Universal Pictures.

Whitt, Laurie Anne and Jennifer Daryl Slack. 1994. Communities, environments and cultural studies. *Cultural studies*, 8(1), 5-31.

Whyte, William H. 1956. *The organization man*. New York, NY: Simon & Schuster.

Williams, Raymond. 1983. *Keywords: A vocabulary of culture and society*, revised edition. New York, NY: Oxford University Press.

Williams, Raymond. 1989. Culture is ordinary. In *Resources of hope: Culture, democracy, socialism*, edited by Robin Gale. New York, NY: Verso, pp. 3-18. (Orig. 1958)

Williams, Rosalind. 2002. *Retooling: A historian confronts technological change*. Cambridge, MA: MIT Press.

Winner, Langdon. 1977. *Autonomous technology: Technics-out-of-control as a theme in political thought*. Cambridge, MA: MIT Press.

Winner, Langdon. 1986. *The whale and the reactor: A search for limits in the age of high technology*. Chicago, IL: University of Chicago Press.

Winner, Langdon. 2003. Testimony to the Committee on Science of the U.S. House of Representatives on The Societal Implications of Nanotechnology, Wednesday, April 9, 2003. Available online: http://www.rpi.edu/%7Ewinner/testimony.htm. Accessed June 17, 2003.

Wise, J. Macgregor. 1997. *Exploring technology and social space*. Thousand Oaks, CA: Sage Publications.

World almanac and book of facts. 2003. New York, NY: WRC Media.

Yamamoto, Traise. 1999. *Masking selves, making subjects: Japanese American women, identity, and the body*. Berkeley, CA: University of California Press.

Index